Conscious Branding

Conscious Branding

David Funk and Anne Marie Levis

Conscious Branding

First published in 2009 by
Business Expert Press, LLC
222 East 46th Street, New York, NY 10017
www.businessexpertpress.com

ISBN-13: 978-1-60649-058-7 (paperback)
ISBN-10: 1-60649-058-3 (paperback)

ISBN-13: 978-1-60649-059-4 (e-book)
ISBN-10: 1-60649-059-1 (e-book)

DOI 10.4128/97816064905

A publication in the Business Expert Press Marketing Strategy collection

Collection ISSN: 2150-9654 (print)
Collection ISSN: 2150-9662 (electronic)

Cover design by Artistic Group—Monroe, NY
Interior design by Scribe, Inc.

First edition: September 2009

10 9 8 7 6 5 4 3 2 1

Printed in the United States of America.

Abstract

Conscious Branding is a step-by-step process that enables entrepreneurs to consciously build a differentiating brand using the Brand Map, a simple visual diagram of how brands function throughout an organization. The Brand Map helps managers and staff visualize their brand and demonstrates how they can make their brand real through everything they do.

Conscious Branding is targeted to anyone who understands the need to differentiate their organization from competitors. The process works equally well for Fortune 500 companies and local service, manufacturing, and retail businesses. The book takes the reader through a linear process, beginning with building a brand platform and then showing how to maintain a consistent brand through marketing communications, product/service design, and operations.

Our goal is to make the complex simple. Type in "branding" or "brand" in any search engine and you'll find tens of millions of pages on the subject. There are scores, if not hundreds, of books on the subject. We don't intend to add to the confusion. *Conscious Branding* describes a simple, easy-to-understand, and tested-in-real-life system. We believe that if one understands the system, or the principles of how brand works, they can make adjustments to fit their particular application.

Our objective with the book is to say as much as we can with as few words as possible. As business people and marketers, we know that most executives already have too much information to process, so we have made the book a quick read, with actionable steps. This is a book born of practical, on-the-ground experience with organizations of all stripes.

Keywords

Brand, branding, brand building, Brand Map, brand foundation, marketing, marketing communications (position, personality, promise), unique brand components, brand experience, brand distribution, identity, naming, TOMA (top of mind awareness), brand structure, market research

Contents

Acknowledgments

We would like to thank Chris Berner and the design team at Funk/Levis & Associates, Inc., for logos and graphics; Claudia Villegas for the development and design of the original Brand Map; Monica Shovlin of the Ulum Group for help and advice in the PR section; Jana Rygas, Anne Marie Mehlum, Amy Gilbert, and Lauren Rathje for their insightful comments on the original manuscript; Kate Watkinson for her help with the operations chapter; Ellen Wojahn for her editing of an early version of the book; and Jennifer Bell for her inspiration and help on the market research section.

INTRODUCTION

Three Objectives

This book is for every organization, whether you make, sell, or distribute a product or provide a service. If you practice what this book preaches, you can increase profits and reduce marketing expenses. This book will help you raise the level of employee engagement and *esprit d'corps* in your business. It will help make your employees proud to work for you. It can assist you to differentiate your company from the competition and put your business at the top of your target customer's mind.

Let's say that you manage a business. You know that job number one, in any business—no matter what you sell—is staying alive. Simply stated, you need to be generating more income than you spend. This is a never-ending quest. Salary, materials, insurance, and all costs continue to rise. You have to constantly bring in more income than the year before. Failing that, you need to reduce expenses. While you are running your business and facing the challenges we have just described, your competitors are doing the same thing.

To survive, you and your competitors either have to steal market share from one another, find new customers, or make more money from the customers that you already have. In each case, you need to connect emotionally with customers and potential customers as efficiently and economically as possible. This is exactly what this book will help you to accomplish.

Think of this book as an operator's manual. Within these pages you will find a step-by-step process to develop and enhance your company's brand at every level of your organization. This process works equally well for large and small companies. We know from experience because we've developed and honed this process with businesses of all types and sizes, from Fortune 500 companies to small startups.

We have three objectives with this book. The first is to demonstrate how to use the brand map and process: a simple visual diagram of how brands function throughout any organization. We have spent over 20 years developing and refining this brand process and the corresponding map. Using the brand map will help you visualize where you are and where you need to focus your efforts. The map is also a useful tool for teaching others in your company about the importance of branding. As you'll soon see, everyone, at every position within your organization, needs to participate in the brand.

Our second objective is to say as much as we can with as few words as possible. We know that if you are taking the time to read this book, you most likely run an organization and are already plenty busy. We are not academicians. We don't need to worry about publishing or perishing. If we were academicians, no doubt this book would be littered with footnotes, and the brand map would have numerous feedback loops and exceptions to the rules. We trust you can figure out any exceptions that apply to your business without our help. This is a book born of practical, on-the-ground experience with organizations of all stripes.

Our third objective is to make the complex simple. Type in "branding" or "brand" in your favorite search engine and you'll find tens of millions of pages on the subject. At our office we have boxes full of books, binders full of articles, and dozens of journals on the subject of brands and branding. There are scores, if not hundreds of books on the subject. We don't intend to add to the confusion. This book describes a simple, easy-to-understand, and tested system. We believe that if you understand the system, or the principles of branding and brands, you can make adjustments to fit your particular needs.

To understand the concept of branding you need to examine a brand in the context of the overall business. The reason to engage in brand building is simple: to increase your bottom line. Numerous studies have shown that strong brands can significantly bring higher prices and contribute to higher profits.

Most business owners are familiar with the essential operational aspects of production, sales, and cash management. However, judging from feedback to Chambers of Commerce across the country, small business owners are less comfortable with the human resource management and marketing sides of their business. These two areas are where effective brand building has the potential to greatly impact a company's performance. This will become evident as you work your way through this book and see how brand is interwoven through the entire fabric of a company.

Building a strong brand is a conscious activity. Rarely does it happen accidentally. To get the most out of this book, you should create your own **brand manual** based on the brand map. We recommend that you create sections based on the suggested brand manual contents found in appendix A of this book. As you move through the exercises and recommended processes of described herein, you can fill in the sections of your brand manual. Now let's get started.

CHAPTER 1

The Importance of Brands

Why do people wear $2,500 Rolex watches when a $40 Timex keeps time just as well? Why did the Toyota Corolla sell for a 25% premium over the virtually identical Chevrolet Geo Prizm, even though both were made in the same factory by the same people? Why do Nike shoes sell for a 70% premium over generic athletic shoes? The answer is simple: branding.

Branded products, services, and companies are worth more and carry more respect than nonbranded ones because people, rightly or wrongly, trust brands. A brand is a combination of differentiating attributes that connect on an emotional level with desired markets. A brand affects the thought process and the emotional responses of an audience. A brand is a relationship. A brand creates value. A brand touches the core emotions and values of its constituents.

When an audience engages your brand in person, through the media, or in other ways, it is called a "brand experience." Ideally, that experience is consistent and controlled by the organization at every point of contact. A brand experience should leave the customer feeling satisfied and eager to—or at least willing to—engage the brand again. Or, in the case of an organization such as a nonprofit, the brand experience should be such that the person who experiences the brand comes away a fan and passes that information on to others. This works at all levels, from potential funders to recipients of the nonprofit's services.

The following examples will give you an idea of what we mean by "brand experience" and how it can easily go awry if you are not concentrating on your brand consistency at every point of contact.

A few years ago, when our marketing communications firm gained a midsize, statewide consumer bank as a client, they informed us that their previous agency had completed a branding program for them. When we asked for a copy of the document so we could determine how to create a communication program for the bank, they showed us a positioning exercise that they had undertaken with the previous agency.

The agency had the bank compare their attributes to a number of other regional banks and determine how they were different from the competition. From this positioning exercise, the agency had created a complete media campaign. It was a nice campaign, but walking into one of the bank's branches, you'd never have recognized that this was the bank that was doing the advertising. The ads had a nice, homey, old-fashioned feel with retro typestyles and grainy, black-and-white photography. The bank branches were chrome, glass, and blond wood. There was a disconnect between the experience of the advertising and the experience of the bank branches. For brands to work, each brand experience has to be consistent at every point of engagement with your customers.

Another example of the disconnect between the brand experience of marketing communications and facilities happened with another one of our clients. This client, a regional coffee company, was developing a drive-through coffee kiosk located in a parking lot on a major thoroughfare between suburbia and the city center. They asked us to design a campaign that would entice the suburbanites to try the kiosk on their way to work. Our plan was to send direct mail in several flights to all the people who lived in the suburban areas that fed into the highway on the way to the city to invite them to drop by and have a free coffee drink. We hoped to turn this trial taste into a daily habit on the way to work. We were somewhat disappointed when the first flight of the promotion didn't draw as many trial tasters as we had expected. To find out why, we went back to some of the recipients of the mailing and asked them why they didn't stop by. Almost unanimously they said the same thing: "We couldn't find the kiosk." The reason? We had mailed a slick black and gold mailer, and the kiosk was colored green. People were looking for a black and gold kiosk. They had driven right by our client. We promptly changed the mailer to match the green kiosk color for the next several flights and the promotion went smoothly because the brand experience of the mailer and the kiosk were the same.

To understand the scope and impact of branding, it is essential to understand that a brand is the result of an aggregate of impressions provided through any number of touch points (points at which your target audience or customers first encounter your brand). This is particularly true in the positive, as it takes many impressions to cement the audience's impression of an organization and its brand. Conversely, it can take just one incongruent or unpleasant experience to weaken or even unravel an organization's image or brand. That one bad experience can impact the customers trust in your brand, which is so important. On the other hand, a consumer with strong brand loyalty often accepts a negative experience as an anomaly and is forgiving if the behavior is not repeated and the negative event is acknowledged and forgiveness asked.

A strong brand

- creates differentiation,
- inspires loyalty,
- inspires trust,
- endures and become memorable,
- creates evangelism, and
- invites press coverage.

Business owners ask themselves all the time, "What makes people buy a product or service from me or, if not from me, from my competitor?" This is the same type of question that political strategists ask when they probe why independent voters will vote for candidate X over candidate Y. In the case of business owners, we first look in all the obvious places: better service, lower prices, faster delivery, more selection, better sales-people, and so on. Political strategists do the same sort of soul-searching: more popular policies, more money, better advertising, more name recognition, and so on. In many cases, those factors are indeed the deciding ones. But one crucial factor that is often overlooked—especially if the competing products, services, or candidates are very similar—is brand. Brand is that overused, misunderstood term that, like weather, everyone talks about, but few do anything about.

So, what is a brand? Our favorite definition for brand is very simple: *A brand is a relationship between an organization, product, or service and a potential buyer.* The key word in this definition is "relationship." Go back to the candidate example. When confronted with two seemingly similar candidates, we will probably vote for the one we most relate to, the one that feels the most like us or the one we are most comfortable with. The same is true for everything from the restaurants we frequent to the grocery store where we shop and the attorney we hire. Surely you have eaten at a restaurant where you weren't comfortable because it didn't "feel" like a place where you belonged. Or you have gone to a store that just didn't carry merchandise that felt like you. When you experience these kinds of situations, it is because you are unable to relate to the company or person providing the brand experience.

We believe that most often people are initially guided by intuition or feeling about something or someone, rather than by logic. When asked why we chose what we did, we find a logical reason for our actions. It's not that we are lying about our motives for choosing one thing over another. It's just that our intuition is quicker than our logic-processing function and, it is difficult to recognize and explain our feelings. The reasons that we "relate" better to one thing over another, or one person, are highly complex and deeply rooted in culture, environment, values, beliefs, and heredity.

The saying that you "can't be all things to all people" is true because we each have such a distinct admixture of values derived from these outside influences that make us all different. It's why there's not just one political point of view or one national tire store with no competitors. And it is a major reason why there never will be a universal political belief or only one option from which to buy your household goods. There will always be an alternative to the Wal-Marts of the world.

It's easier to understand the importance of relationships to political candidates because it is relatively easy to understand the concept of one person relating to another. It is a bit more difficult to understand people relating to products or services except beyond the direct relationship with the salesperson. But if you stop and think about it, we do relate to things all the time. Ask any dog lover. Or coffee drinker. Or Chevy lover. It also turns out that because of culture, age, and physiology, we relate to certain colors better than others. And as you'll see later in this book in the section discussing logos, we even ascribe different meanings to different shapes.

The secret to building a brand is to understand yourself and your market and use the tools of communication and branding to build a relationship with your customers. Sounds simple, huh? We think it is.

CHAPTER 2

Brand Confusion

"Oh, yeah," the plant manager said as we toured his site. "We had our company brand done by our first ad agency. That was years ago. We knew branding was important. You can't help but read about it in our trade magazines. Branding is the buzzword in our industry. Didn't you notice our brand? It's on our business cards, our building sign, and our trucks."

We've engaged in a variation of this conversation more times than we can count over the past two decades. People still confuse brands with logos. Even though the discussion of brands and how they work has been around for half a century, brands are still a mystery to many business people.

Brands, in a more primitive form, have been around a very long time. We marketers just never got around to understanding them and how they work until relatively recently. The truth is, we are still learning about branding all the time. Early intentional branding efforts were little more than advertising to establish a brand name. In an article in the June 2006 issue of *Discover* magazine, author Heather Pringle describes the 2,000-year-old "graffiti" of ancient Rome. This graffiti was different in style from what we see today. Back then, the graffiti artists scratched their graffiti into plaster walls or even hired professional stonecutters to engrave their thoughts on walls and tombs. In these ancient inscriptions we find references to a particularly well-known and cherished wine called Falernian. We also find evidence of product branding in baking molds used to make bread and cakes for sale at gladiatorial contests. One mold reads, "Miscenius Amphilatus makes [this] in Salonae."

So brand efforts existed quietly for centuries until certain observers began to notice that people had relationships with products and organizations that went beyond the basic fulfillment of easily observable needs like hunger, shelter, and feeding of habits. Initially when this realization dawned, the discussion of the concept of brands was confined to academicians and business theorists. It took a while for the term "brand" to slip into the consciousness of the average businessperson and when it did, suddenly it seemed to be everywhere. Every trade magazine, no matter the industry, seemed to have an article on branding. The Internet, with its far reach and fast distribution, was filled with brand articles and Web sites. It was a case of too much information. This plethora of articles, rather than clarifying the idea of brand building, just confused the concept further.

There is another factor that affects people's lack of understanding about brands. Unfortunately, when people first hear the word "brand" or "branding," they think of the ownership symbols burnt onto the flanks of cattle. Thus, brands easily became confused with logos and identity systems. That confusion remains persistent and ubiquitous to this day. Hence the reason we seem to have the same conversation—like we did with the plant manager—over and over again.

There seems to be no unanimity about what a brand is and how it works. Even in our field, when we attend brand seminars or read trade articles, we find confusion about what a "brand" is and how you attain it. Some marketers think branding is the same thing as positioning. Positioning, in our view, is actually just one of the legs of the three-legged brand stool. We'll discuss positioning in more detail in chapter 5.

CHAPTER 3

The Three Lessons of Conscious Branding

Before we get to the heart of the brand building process, allow us to give you three examples and demonstrate the branding lessons we can all learn from them.

Lesson One

Let us say that you are about to open a new business. You are relatively new to town and you've asked around for references for a good local law firm. You've heard several names and from that you have narrowed it down to two firms, one because it's close, the other because the name has come up more than a few times. You look them up in the Yellow Pages to get their phone numbers. Their Yellow Pages ads look quite different. Firm A's ad lists their name and their services and has a big, bold phone number in block letters. Firm B's ad has their name, a tagline about serving the needs of business since 1976, a list of services, a phone number to call for further information, and a Web site. The latter's ad uses a consistent typeface (Century) and stands out because of the generous use of white space.

You check out the firms' Web sites and see that Firm A has a nice group picture of well-dressed attorneys, lists the specialties of the firm, and shows a map to the office location. Firm B's site looks a lot like their ad; it starts with the firm's philosophy, shows individual pictures with resumes and specialties of each of the firm's attorneys, and shows all the employees, including receptionists and legal secretaries. You withhold judgment and decide to interview the two firms.

You make a call to each firm. When you set the first appointment, Firm A's receptionist is crisp and efficient as she turns you over to another scheduler for an appointment. Firm B's receptionist is friendly and takes her time to understand what you need before connecting you to an attorney who greets you, asks a few initial questions, and then sets an appointment.

You arrive at the Firm A's address, a detached brick building with an asphalt-surfaced parking lot and a monument sign leading you to the front entrance. The sign is made of backlit white plastic with blue Helvetica lettering, encased in an aluminum frame, and set in a small lawn. A few small weeds and wind-blown paper nestle at the bottom of the sign. The cement walkway shows the effects of decades of weather and foot traffic. You walk up the cement steps to a porch and a glass door with the firm's name in gold applied lettering not quite matching the sign in front. Upon entering the lobby, you gaze around for a second to determine where to go next.

The lobby is a rectangular room with a built-in reception desk opposite the entrance. Two pairs of leather swivel chairs are separated by a small oak table with law magazines, all set beneath the firm's name in raised black Times Roman letters on the white wall. Against the opposite wall is a modern green couch with an equally modern black lacquered end table, set with a small vase of convincing fake flowers and a thin, modern lamp. The floor of the lobby is carpeted in a patterned gray and black carpet. The ceiling—of normal height—is swirled, white plaster.

Behind the reception desk, a young woman on the phone acknowledges you with a nod of the head and a slight smile before returning to her conversation. Not knowing whether to stand there or sit, you move to one of the swivel chairs to give her some privacy for her call. With an "Okay, talk to you later," she hangs up and looks your way as you are settling into a chair. "Can I help you?" she asks in a professional tone. You stand from your chair and tell her who you are there to see. She tells you to take a seat, and she calls the attorney. She informs you that his office is the third door to the right as she gestures down the hallway behind her. You stand again and proceed down the hall. You look in the open third door. The lawyer looks up from his desk and stands to greet you.

The attorney is well dressed in a dark suit. He greets you warmly and asks you to take a seat opposite him at his desk. He offers you coffee. You accept, and he buzzes the receptionist to bring it to him. She delivers the coffee wordlessly, and he thanks her with a nod of his head. You then proceed with the interview. After half an hour, with all your questions answered, he escorts you past the receptionist and into the lobby, holds open the door, and bids you goodbye.

Now let's look at the experience with Firm B. You arrive at the law firm's address, an eight-story building with underground parking. An elevator takes you to the lobby of the building. The firm's name is on a building directory in the large stone and marble building lobby. The sign directs you to the seventh floor. You enter a well-lit elevator with the same feel as the lobby and ascend to the seventh floor. As you exit the elevator, a brass sign on the wall in front of you directs you to the left for the offices of the attorneys. You look to the left and see the tall, wide, cherry double doors opened to a beautiful cherry reception desk with a brass reveal. On a four-foot-wide, floor-to-ceiling cherry panel behind the reception desk is the firm's name in brass letters, in the same Century typestyle as the elevator lobby sign and the Yellow Pages ad.

The lobby of the law office has sweeping curves. The 15-foot-high ceiling is punctuated with recessed lights casting a warm glow over the space. The whole effect is one of prosperity and good taste. The furniture is inviting and modern, the carpeting a subtle pattern of warm tones.

Behind the low reception desk, a well-dressed young woman smiles brightly as she directs a phone caller to the proper attorney. She stays engaged with you, smiling with her eyes as she speaks on the phone. When she hangs up, she asks how she can help you. When you explain that you are there for an appointment, she calls an extension and says that the attorney will be right with you. She offers you a seat and a cup of coffee or tea or a glass of water and inquires about your day. You accept the coffee, with milk, and she brings you a cup on a saucer. As you sit there, she engages you in conversation until interrupted by the phone. She excuses herself and answers it.

A well-dressed attorney enters the lobby and walks toward you, hand outstretched with a warm smile of greeting. He addresses you by name and invites you to a nearby small conference room. On his way, he thanks the receptionist for getting the coffee. In the conference room, he sits next to you, turns his chair your way, and asks about you, your business, your family, and how you fared the recent storm. After a few minutes of conversation, you proceed with the interview. After half an hour, with all your questions answered, he escorts you past the receptionist, who thanks you by name for visiting and says she hopes to see you again soon. The attorney walks with you to the elevator, pushes the ground floor button for you, and thanks you for considering their firm.

By now you certainly see where we are going with this example of two firms in the same line of work. Between the first impression created by the Yellow Page ads and the time you left each office, you have picked up dozens of conscious and unconscious cues about each organization. Even though you are making a business decision in deciding which to hire, your decision will be greatly influenced by how these two firms made you feel about them and about you. The experiences at each firm set up your expectations on all sorts of levels. Which firm do you think is the more expensive? Which firm do you think is the more successful? Which firm are you more comfortable with? Which firm do you think will take better care of your legal needs? Which firm do you trust more?

In these cases, the experience began with the word-of-mouth recommendations, followed by the ads in the Yellow Pages, then by the Web sites and calls to the receptionists, and finally, to the person that the receptionist connected you with. Already, before you even stepped into the offices, you had gathered a host of cues. Upon arriving at and entering the offices, being greeted, seeing the office space, and talking to the people there, you gathered still more information. Much, if not most, of the information you collected was nonverbal and not expressly what we would typically call "marketing communications." Yet all of these cases communicated at a much deeper, emotional level than paid media usually accomplishes.

This is the first lesson of conscious branding. *A brand is a total experience.* It is not a single event. It is a series of events. These events combine to create expectations of a relationship with a brand. This leads us to lesson two.

Lesson Two

One cruel, winter evening my wife and I were traveling slowly and carefully over the Cascade Mountains in Oregon in the middle of a driving blizzard. As we approached the 5,000-foot Santiam Pass, we were slowed by warning flares and waved over to the side of the road by a state trooper who stood in the howling storm and shouted that we couldn't proceed any farther without chains on our tires. We pulled over to the side of the road, joining dozens of other cars with blinking warning lights, and prepared to brave the harsh storm. I pulled on my winter jacket, donned my wool cap and heavy boots, and plunged out into the maelstrom. Stinging snow pelted my exposed cheeks as I pulled the canvas chain bag from the rear of my four-wheel-drive vehicle. The eerie glow of snow-muffled headlights and warning blinkers flashing against the wind-driven snow created a foreboding scene as I struggled with the flashlight in my gloved hands, trying to read the proper way to apply the chains. Finally, cold and stiff, I pulled the tangled chains out of the bag, knelt in the snow- and ice-covered pavement, put the flashlight in the snow next to me, and, shivering from the freezing wind, laid one set of chains next to my wheel.

Just then, a figure materialized out of the darkness. It was a thirty-something man wearing a blue jacket with an insignia. The insignia said, "Les Schwab Tire Centers."

"Let me give you a hand with those chains," he said. "I'm pretty fast with them. Do you mind holding the flashlight?" Before I could reply, he handed me the flashlight, bent down, and took over, installing both chains in a few minutes.

"How much do I owe you?" I asked. "Not a thing," he replied. "I was just driving over the pass myself, and it looked like a lot of folks needed a hand." With that, he bid me goodbye and moved to the car behind me.

Les Schwab is a tire company in the western United States that has built a brand known far and wide for its exceptional service. Les Schwab Tire Centers was founded in 1952 in Prineville, a small desert town in Central Oregon. The company founder and namesake, Les Schwab, started with the purchase of Prineville's OK Rubber Welders Tire Store and expanded it into nearly 400 stores in the West.

Les Schwab Tire Centers employees are masters at not only creating expectations but exceeding them. When you pull up to one of their stores, an employee runs—not walks—out to greet you. He or she finds out what you need and immediately gets you to the right person if they can't help you. If you are there to purchase tires, they don't try to sell you the most expensive set. Rather, they try to understand your needs and expectations of a set of tires. Then they try to match the right tire to those needs. If you are having tires installed, they try to get to them instantly, in which case they escort you to a comfortable waiting area, or they promise to have them ready by a certain time. They are always ready ahead of that time.

Les Schwab also will fix any punctured tire you bring to them for free, regardless of whether you bought it there or not. They also have a policy that if you buy your winter tire chains from them and don't use them over the winter, you can bring them back and get your full purchase price refunded. Both of these examples are great instances of exceeding the expectations you would probably have of a tire store. My winter chain installation example sets a standard that is hard to beat.

It is our belief that people develop expectations from initial experiences. In general, when these expectations are met or exceeded, the beginnings of a relationship are formed. Continuously meeting or exceeding the expectations cements that relationship. This is true in the positive and the negative. In other words, if someone has expectations of a negative engagement with your organization because of a past negative experience with it, or even a bad impression from an ad or word of mouth, they will be expecting you to fulfill these negative expectations. They'll come in with an attitude. Certainly everyone has experienced going into an office with the expectation of poor service. Your typical Department of Motor Vehicles (DMV) comes to mind. In fact, when the local DMV turns out to be a not-so-bad experience because of a nice clerk or short lines, you happily accept service that would annoy you in a different setting.

If you have built a strong relationship with a customer, generally an occasional lapse, if properly corrected, won't be too injurious to your brand. The beginning of any relationship is the most fragile, so it pays to plan the customer's initial experience with you very carefully.

Relationships depend on trust above all. People buy certain brands because they trust them to meet expectations. Exceeding expectations, especially in ways a customer wouldn't request or anticipate, can cement relationships so that they are almost unbreakable. Les Schwab Tire Centers is a perfect example of that.

This is the second lesson of conscious branding. *A brand is a relationship built on a foundation of expectation and trust.*

Lesson Three

In 1995, Samsung Electronics, a South Korean company that made memory chips and other components that powered the high-tech products made by brand name Japanese and American companies, had a pretty good year. Though Samsung had their own branded line of inexpensive commodity consumer electronics, their bread and butter was their component business. At the end of 1995, the chairman and son of the founder, Kun-Hee Lee, gave Samsung mobile phones as gifts to friends and key executives. Within days he started receiving complaints that the phones were defective.

Embarrassed, he ordered that the entire $50 million inventory from their Gumi facility be piled in a giant heap in the factory courtyard under banners proclaiming that from that day forward Samsung would be known as a maker of quality goods. Lee, his top managers, and 2,000 employees watched as workers smashed 150,000 phones and fax machines and threw them into a roaring bonfire.

Although it was likely a highly toxic bonfire, it certainly made a point. Brands not only can change, but they must change. Times change, styles change, technology changes, and markets age and evolve. Brands, in order to survive, must change with the times. Rarely are changes as dramatic as Samsung's.

The second point made from this dramatic bonfire display was the message that the change in corporate direction and the building of the new Samsung brand came from the top of the organization. The dramatic display and subsequent change in direction paid off. In 2006, Samsung surpassed Sony in revenue. This is just another real example of the power of deep and conscious branding.

Brand building is a nonstop endeavor. It is not a one-time exercise or a manual on a shelf. A brand is a living, breathing reality. This is the third lesson of conscious branding. *A brand is a journey, not a destination.*

Keep these lessons in mind as you begin to build your brand:

1. A brand is a total experience.
2. A brand is a relationship built on a foundation of expectation and trust.
3. A brand is a journey, not a destination.

Keeping these three maxims in mind is critical to brand success. Look at your organization through the eyes of the consumer. Try to experience the entire process of brand engagement as described in the two earlier law firm scenarios by examining your brand from the very first point of contact. What do you need to adjust along the path of engagement to ensure that the customer experience matches the expectations that you have of your brand? Talking to customers yourself, or better yet, hiring competent researchers to do it for you, can yield surprising and useful information to help you fine tune your brand experience.

Try to learn what your customers expect of your industry in general. Brainstorm ideas about how to exceed those expectations. Build in surprise and delight. Pay attention to the smallest details. Apple, Inc., does an excellent job of surprising the consumer along the way. Ask anyone who bought one of the original Apple iPhones about the packaging. Just opening the box was a delight. Apple's designers designed an entire experience that began at the store and continued through the packaging and through each new software update. As a result, Apple aficionados are fiercely loyal to the brand and trust and expect Apple to "wow" them and provide functional, easy-to-use products.

The Brand Team

In the early nineties, when we were first developing the branding process on which this book is based, we were dismayed to find that all to often the participants in the process were excited and fired up to improve their brand when they first completed the exercises, but soon the brand manuals were languishing on the shelves and people were back to their old habits. We understood the problem. Everyone is too busy, change is difficult, and the daily battles seem to take precedence over long-term change. So, as we were fine-tuning the brand exercises, we were also experimenting with ways to keep change alive. We learned that brand building takes both leadership and commitment from the top and a focused brand team that meets regularly. The team should have defined goals and measurable objectives, with responsibilities with completion dates assigned to specific people.

Your brand team should consist of people from throughout your organization. Branding is the responsibility of the entire organization, not just management or the marketing department. The brand team should meet regularly, at least quarterly, to develop actionable recommendations for strengthening the brand and assessing the success of earlier branding initiatives. Leadership of this team doesn't necessarily have to be a company executive, but it should be someone who is enthusiastic about the company.

CHAPTER 4

The Context of Brand

Sometimes, the day-to-day job of running a business can seem like trying to shovel smoke with a pitchfork in the wind. When each day is an effort just to maintain control in changing conditions, it's easy to put off brand building with the excuse that you have too many other fires to put out. But it needn't be that difficult a chore. Brand building should be fun and enlightening. It should give you a clearer sense of what your task is and how to get it done. It should help you improve your bottom line.

There is a reason why people don't start businesses and why most business start-ups fail. Business is exceedingly complex and difficult. A business needs vision, capital, organization, a product or service, marketing, sales, and relevance. We maintain that to be successful in the long term, a business or organization needs a strong brand that is woven into all aspects of running that business. Branding is such a part of the other operational aspects of a business that to ignore the influence of brand on operations, and operations on brand, is to put your long-term organizational survival in jeopardy.

It is beyond the scope of this book to be a primer on business, but since brand is part and parcel of business success, it is reasonable to look at how brands fit into the overall business system.

Vision

In our experience, many young organizations are strong on vision and weak on planning. Conversely, many large organizations are strong on planning but lack clear vision throughout the ranks. Vision is woven into the brand of a company. In fact, it is considered by some to be the basic foundation of a brand. Vision describes where you want to go as an organization and what it might be like when you get there. Vision is an important consideration in brand planning. Brands should be aspirational. They should reflect what your vision is as much as, or more than, who you are at present.

Capital

A strong brand makes the generation of capital easier. Lenders are more comfortable with businesses that appear successful and customers are more comfortable buying branded products. Building a brand is the process of consciously aligning symbols that represent who and what you are. You may have all the other aspects of your business running smoothly, but if that's not discernable to people on the outside, you are just making it tougher on yourself to succeed. Branded products have been shown in study after study to generate more profits than unbranded products.

Organization

Organization means that you need to have a structure of operation that enhances the means of generating capital and of acquiring and retaining employees and customers. Within the category of organization are operations, human resources, production, and the like. As you will see in future chapters, inculcating the brand deep into your company structure and organization is essential to building an enduring and successful brand. The most difficult concept for many organizations to grasp is that building a strong brand is a function of not only the marketing department but of every department throughout the organization/business.

Marketing/Sales of Products and Services

Marketing communications and direct sales, both components of the larger category of marketing, are the most commonly thought of ways to promote your brand. However, the products or services that you offer are also perfect vehicles for carrying the message of your brand. In some cases your entire company brand is dependent on your branded products and services. These are often the only way that people engage your company and are what generate your capital, either directly or indirectly. If the consumer's first experience with one of your branded products and services is a positive one, it is much easier to sell them a second product.

Banks, for instance, rely on this type of cross selling to a great degree. They hope that a good experience with a certificate of deposit or a home loan will lead to a checking account, which might lead to a consumer loan, and so on. As we move through this book you'll see how to carry your brand through to your products and services.

Relevance

Relevance means that your organization is serving a purpose outside of its own maintenance. That purpose is generally that you are somehow enhancing people's lives. Relevance also means that you, or your products and services, matter to people, that you are differentiated from similar organizations that serve your target population, and that your differentiation is a reason to engage with your organization. In other words, people have to care about you.

There are a lot of reasons to put effort toward consciously building your organizational brand. Here are a few:

- People will buy extended products from a brand they trust.
- Customers will cut you slack for your occasional missteps if they have a relationship with you.
- Branded products have better margins and reduced marketing costs.
- Brands make it easier for customers because the customer doesn't have to put as much thinking into product or service selection.
- A strong brand makes it easier to recruit quality staff.
- A brand differentiates you from competitors.
- A brand promises authenticity.
- A brand increases memorability.

In the competitive world of business, the best brand wins.

CHAPTER 5

The Brand Map

The brand map (see Figure 5.1) was designed to show at a glance how every part of your organization is connected to the process of building and maintaining a strong brand. It will help you visualize how all the areas of your company are connected and can work together to reinforce the brand experience. The brand map will also help you organize how you will promote the brand at every touch point in your organization.

Your own brand map might differ slightly from our example, based on the nature of your business. In addition, most of the business activities we show on this map extend out in greater detail. We only demonstrate this greater extension in the "logo" example, even though most every example has more depth and nuance. We will explore the brand map in more detail in upcoming chapters.

The basis of the brand map is the brand foundation. It is from this foundation that the brand is developed throughout the organization, so it is important to spend time building a strong foundation for your brand.

One of the things that you will discover if you study brands is that there are many definitions of what a brand is and what the components of a successful brand include. As we mentioned in chapter 1, our definition of a brand is a relationship. But how do you consciously create that relationship? To keep things simple, we have narrowed down the key components of a brand into a simple brand foundation that consists of the three P's:

- Personality
- Position
- Promise

For the purpose of integrating a brand foundation into the very essence of an organization, it's our belief that position, personality, and

BRAND MAP

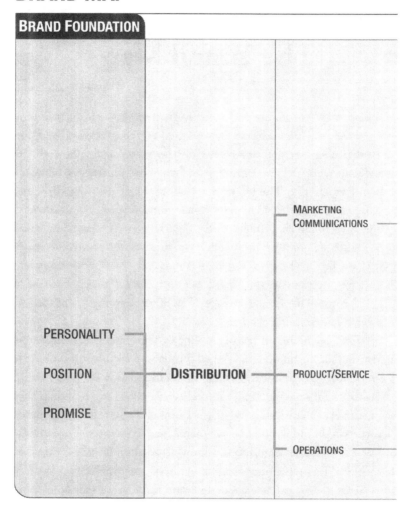

Figure 5.1. Brand Map

IDENTITY
- NAMING
- LOGO — SILHOUETTE
- COLOR — GESTALT
- GRID
- TYPOGRAPHY
- TAGLINE
- ICONOGRAPHY

ADVERTISING
- RADIO
- TELEVISION
- PRINT
- SOCIAL MEDIA
- OUT-OF-HOME
- WEBSITE

COLLATERAL
- BROCHURES
- SPECIALTY

PUBLIC RELATIONS
- PUBLIC SPEAKING
- MEDIA RELATIONS
- KEY MESSAGES
- EVENTS

COMMUNITY INVOLVEMENT
- CHARITABLE GIVING
- SPONSORSHIPS
- INDUSTRY ASSOCIATIONS
- COMMUNITY BOARDS

STYLE
- TONE
- KEY WORDS
- VOICE

TRADESHOWS
- DESIGN
- TEAM TRAINING

PRODUCT DESIGN

PACKAGING

CO-MARKETING

BRAND EXTENSIONS

FACILITIES

TRAINING

PROCEDURES
- COMPLAINT HANDLING
- NEW CLIENT INTAKE
- SALES PROCESS
- CUSTOMER EXPERIENCE

HIRING

promise together are the brand foundation elements that most easily translate to all touch points. Ideally, each touch point, or point of contact with your organization, would reflect your company personality, position, and promise.

Our definitions for each of the three elements are below. Later we will go into detail about how to determine your own personality, position, and promise.

Personality: The "Who" of Your Brand

Personality is comprised of the human-like traits of an organization. Traits such as warm, fun, professional, reliable, serious, energetic, aggressive, and so on can both help differentiate an organization from its competitors and establish the basis for a relationship between an organization and its constituencies. Your personality determines how you speak, your tone of voice, and the words you use. For example, a young, brash personality would not be appropriate for a company that markets to senior citizens.

Position: The "What" of Your Brand

Position represents the distinct point of differentiation that separates an organization from its competitors. Position is comparative and exclusive. In other words, if you hold a position like "the low price leader," comparative competitors don't, by inference, hold that position also. The position is exclusively yours. Position is relative to competition and usually refers to leadership in a particular attribute. For an excellent primer on the concept of positioning, check out the easy-to-read book *Positioning: The Battle for Your Mind* by Jack Trout and Al Reis.

Your position (and your personality and promise) may also be represented by your company history, your people, your unique products and services, or your delivery methods. Every organization has a collection of unique qualities that differentiate it from competing organizations. While an organization may share similar qualities, such as friendliness or speedy delivery with competitors, specific differentiating qualities, which we refer to as unique brand components (UBCs), separate you from any other organization. UBCs are a part of your position. Later we will examine how to determine these UBCs.

Promise: The "How" of Your Brand

The "brand promise" refers to what people can consistently expect from your organization. It is the standard that you will always do your best to achieve. Usually this refers to an experience they can expect. For example, McDonald's promises the same taste and style of food at any of its locations. Starbucks' brand promise is a consistent welcoming experience, rich coffee smells, and high quality products.

The brand promise also considers and incorporates your values and mission. For instance, if a core company value is that customers are always right, then one actionable aspect of your promise might be that you make it easy for them to be "right" by empowering employees to make decisions on the spot rather than making the customer work their way through management to get satisfaction. The brand promise is the commitment you make to your customers and an important component of differentiation.

A great example of empowering employees to keep a promise that relates to values is found in another Oregon-based company—Dutch Bros. Coffee. Dutch Bros. is a chain of drive-through coffee stands, recognizable throughout Oregon, Washington, Idaho, California, Arizona, Nevada, and Colorado by their Dutch-themed blue stands. They are also known for having a very loyal following. The founders of Dutch Bros., brothers Dane and Travis Boersma, believe that building a relationship is a one-on-one endeavor. Their employees on the front line, the ones that make and serve the coffee drinks, are encouraged to get to know their customers by conversing with them. They have the freedom to make decisions that will help build those relationships. For instance, if a barista discovers in conversation that the customer is having a bad day, or has gotten a promotion in their job, the barista can offer a complimentary coffee drink to cheer them up or congratulate them. This unexpected gift cements loyalty and is well worth the small cost. The word-of-mouth advertising is priceless. The founders of Dutch Bros. see coffee not as a product but as a service that reflects their personal values of treating everyone with decency and respect.

Once an organization has defined each of the three elements of their brand foundation, the next step is to "distribute" those key elements through all touch points. Touch points are sometimes referred to as "points of contact," or every single place where the public comes into contact with your organization. Every touch point should reflect the brand foundation. As you can see from the brand map, there are a myriad of touch points for any brand.

CHAPTER 6

Building the Brand Foundation

Now that you have some background, let's get to work on building *your* brand. In this chapter we will take you through the process of building your brand foundation. We recommend that you start a binder or a file in your computer that codifies your brand as you go through the steps of building it. Ultimately, you should have addressed every distribution point on the map, or at least have a clear idea of what you need to work on. In this and subsequent chapters we will indicate specific actions you can take under the heading "Here's How."

Defining Your Organizational or Product Personality

We've said that building a brand is similar to developing a relationship. In the case of branding, it is building a relationship with your customers, vendors, and employees. Building relationships is a primal instinct in social animals like humans. We build relationships and loyalties with family, friends, communities, and nations. We like our football team and can't stand the rival. We might like the Republicans and not the Democrats. We even build relationships with cars, restaurants, and animals. We relate to our fellow mammals—like dogs and cats—because we can attribute human-like characteristics to them. We don't relate to goldfish in the same way. We relate to personalities that seem human and complement the kind of personality we have as individuals. We also relate to the kinds of personalities that we admire or aspire to have ourselves.

Brand personality is critically important. It is common marketing knowledge that people purchase products that reflect and validate their values. This phenomenon is most obvious in consumer goods such as automobiles, watches, and cigarettes. Marlboro sold rugged manliness to men who felt they fit—or wanted to fit—that image. Harley-Davidson sells the outlaw image to those who feel it fits them or would like such an image granted to them by association with the brand. Seiko watches are marketed not only on the basis of time-keeping accuracy but also on status conferred to the owner by the brand.

Think about yourself. If you were to go out and purchase a new automobile and you felt you had enough money to get any car you wanted, what would you look at first? We'd wager that you would first consider cars that fit your values and your self-image before you sat down and compared horsepower and mechanical specifications. This is because cars have moved beyond mere transportation machines to another level. Cars now are outward manifestations of who we are. They are extensions of ourselves. The same is true with many products and services. It is why Versace, Coach, and Gucci sell for much more than non-brand-name products in their same categories.

Brand personality research has been going on for decades in academia. Some, such as studies by Jennifer L. Aaker of Stanford University's Anderson School of Management, have focused on five dimensions of personality—sincerity, excitement, competence, sophistication, and ruggedness. From these five categories, Ms. Aaker suggests a brand personality framework that includes 42 personality traits that she calls the brand personality scale. This is not to say that there are only 42 personality traits. There are actually hundreds of possible adjectives to describe personality.

To demonstrate how personality is imbued in colas, Ms. Aaker cites examples from earlier studies that show Coke as cool, all-American, and real; Pepsi as young, exciting, and hip; and Dr. Pepper as nonconforming, unique, and fun. She also describes how products have gender. For example, she says that Virginia Slims cigarettes are perceived as feminine and Marlboro cigarettes are perceived as masculine.

Personality is perhaps the easiest element of the brand foundation to determine. Like position, it is best determined by outside research, but it can be easily done internally first, then verified by qualitative and quantitative research. Or, if budget prohibits extensive outside research, you can poll internally and verify with a few customers to get a fair representation of your organization's personality.

Here's How

Personality, as we said earlier, is comprised of human-like attributes. To describe your organization's personality use the same kinds of words that you would use to describe a person.

Personality Exercise 1: The Words to Describe Us

Gather your brand team to brainstorm and narrow down adjectives. Are you professional? Fun? Aggressive? Smart? Concerned? Ask the brand team to choose some adjectives to describe the personality of the organization. Don't veer off into words or phrases that wouldn't be used to describe the personality of a person. Build a list. Look for common themes. Then pare the list down to five or so salient personality attributes. You can either do this as a group or with a smaller brand team. When you are finished paring it down, bring it back to the whole company to get feedback. Total employee engagement is important for three reasons. First, it allows everyone to participate in the brand and gives each one of them ownership. Second, it helps the employees see what personality attributes they should be representing. Third, it is important to see your brand holistically and at all points of the customer relationship.

This process is simple, but beware. Internal self-analysis can be myopic. Often, especially in a brand team situation, one strong personality can affect what the group thinks. When we perform this exercise with companies, we often ask the president to speak last so that he or she doesn't unduly influence subordinates. In any case, once you have narrowed your personality attribute list down to a manageable handful, check with a representative sample of current customers, past customers, and noncustomers. What adjectives would they use to describe your organization? If their lists are different than yours, you may want to undertake further research or accept that your personality adjectives are, as yet, aspirational. If you choose the latter, you will need to institute training programs to help your organization develop the desired personality. You will also need to tailor your marketing communications program to express the new desired personality attributes.

Personality Exercise 2: Our Company Representative

Imagine your company representative. In performing this exercise, it is very helpful to imagine what, if your company or product were a human, he or she would look and act like. Think of a public figure such as an actor, a politician, or a person from history that feels like your company. Another way to determine your brand personality is to ask, "If we were doing a TV advertising program, who would we choose to be our company spokesperson?" Once you have determined who that person is, try to figure out what personality traits they project. These will help you to build your list of personality traits.

You can also look for pictures in magazines or in stock photography for what we call your "representative icon." Keeping a mental image of this icon can help all the members of your organization to stay on track with a consistent portrayal of your personality. Once you feel that you have successfully defined your personality and found your representative icon, you should record what that personality is in as few words as possible. One way to do this is to create a single sheet that has a picture of your representative icon with a brief paragraph describing your personality. If you are really concise, you might be able to accomplish this personality description with simple bullet points.

Personality Exercise 3: If Our Company Was an Animal, What Animal Would It Be?

Sometimes the most easily imaginable and strongest company or product personality icon is not a person but an animal—like the gorilla of Gorilla Glue, the friendly cartoon airplane for Expedia, or the Rock of Gibraltar symbol for Prudential insurance company. These personality icons work because they already have cultural connotations in our world. Everyone knows that gorillas are strong, fat cartoon characters are friendly, and rocks are solid. Think of an inanimate object that represents your brand. Or, have your brand team think of your organization as an animal. What is that animal? Why? How do the animal traits relate to your company's personality?

After you have conducted the three previous exercises (or any combination of them), develop a list of five (or so) personality traits. When you have finalized your company's personality attributes, record them in your brand manual. Note that this can be done for products as well.

Here's an example of how we helped a company determine their product personalities using a representative icon for each product. Our client produced creams and lotions that were designed to prevent and cure poison oak and poison ivy. Their flagship product had been on the market for several decades. They were under increasing competition and their market share was dropping because the category was getting saturated with "me too" products. To build more sales in the category and compete with the newcomers, they had developed an alternative formula that had different properties than the original product. Our objective was to introduce a new product in a way that didn't cannibalize their flagship product but capitalized on that product's long-standing reputation and name familiarity. Our most important objective was to regain market share from competitors.

Before we started on the new product, we revisited their flagship product. It had been on the market for almost five decades, so it had credibility, yet also appeared old fashioned. We needed to find a way to take advantage of the positive aspects of this time-tested formula as we moved into a new formulation and product. We discussed with the client what words might be used to describe the product if it were a person rather than a product. The words that came out for the original product were reliable, trusted, consistent, not flashy, dependable, and respected. We also determined that the product was male and middle-aged. Armed with this information, we looked through outdoor magazines and other sources for pictures of people that our description might fit. When we came across the picture of a National Park Ranger, square-jawed, looking directly at the camera and wearing his crisp brown uniform with the "Smokey the Bear" hat, we knew immediately that we'd found our representative icon for the flagship product.

Then we tried to determine how such an icon would talk. What words would he use that would be consistent with who he was? We developed a list of words and a precise style of copywriting based on our determinations. We wrote all of this information down on the sheet where we'd pasted his picture. Our park ranger became the product personality for the original product. We even used a picture of a square-jawed park ranger in our advertising and promotional literature.

We repeated this process for the new product. This new product was a bit more expensive and was designed to be carried easily in a backpack or pocket. Unlike the flagship product, which came in a plastic bottle, our new product came in a plastic tube. It was designed to be used by active outdoors people. The product had scrubbing beads that contrasted with the smooth, creamy texture of the flagship product. The packaging we developed was aimed at a younger demographic and utilized bold, graffiti-style type; bright red color blocks; and gold foil. The words that we used to describe this product personality were aggressive, bold, extreme, fast, active, and tough. From these words we determined that the product was male and in its late twenties. The representative icon we came up with was a handsome, young mountain biker carrying his bike over his shoulder. In our picture, he was shirtless, lean, and muscular. As before, we determined how he would talk and the words he would use. We also used a model to represent him in our advertising and promotional materials.

Developing a representative icon may or may not work for you, but the idea is worth considering. The representative icon "transfers" a perceived personality to a product or company. That is the reason companies use spokespeople and even cartoon characters to represent them. The bulbous, protective Michelin Man, born of a pile of tires at the beginning of the twentieth century; Mr. Peanut, with his snappy, man-about-town persona; and other characters abound in the business world. Nike is famous for its use of well-known, distinct sports figures with big personalities to represent Nike's in-your-face brand personality. Using an icon, even just as an internal tool, can help your organization take your brand to the next level.

Determining Your Company Position

Since the publication of *Positioning: The Battle for Your Mind* by Trout and Reis in 1981, thousands of articles have been written on the subject of positioning. The book has become a classic in marketing literature and has revolutionized the marketing world. Yet, the premise is simple. So simple that it intuitively rings true. In brief, the authors contend that, as humans, we naturally compare and rank things. Being first, or number one, is best. Being number two isn't as good as number one but is competitive. The authors offer examples and strategies for how to position a company and how to market when you are number one or two.

Being number one doesn't necessarily mean being the biggest. A company can also be number one in low price offerings, in speed of service, in quality, and in any number of other attributes. Being number one means holding a leadership position. When you hold a leadership position, people feel more confident about their buying decision. It's always safer, the thinking goes, to buy a leading brand.

At the turn of the twenty-first century, in the world of automobiles, Volvo held the leadership position in safety. Toyota held leadership in reliability. Lexus and Acura fought it out for affordable luxury. Rolls Royce has a long held position of exclusivity and handcrafted luxury. Mercedes Benz stood for quality. Kia and Hyundai were the low-cost leaders. These positions can and do shift over time as auto manufacturers adjust to changes in the marketplace. Oftentimes you'll see that when companies allow their positions to erode, so do their reputations and their sales. Cadillac used to be a leader in affordable luxury. Now it is trying to reestablish leadership in styling.

How do you figure out your position? Determining where your organization is positioned is a simple matter of outside research, or if budget is a limitation, you can do a pretty fair job of determining your position with a small group of in-house and customer participants. In either case, you start the same way, by determining what attributes are important to consumers. You can use qualitative research like focus groups to determine what these attributes are, or you can use your best judgment based on your experience in the business. The following are two ways you can approach determining your position internally.

Positioning Exercise 1: The Positioning Matrix Process

Positioning works for products, services, and companies. You can do a fairly simple positioning exercise yourself. If you want to do this in-house, try this Positioning Matrix exercise.

Here's How

To begin, on the left-hand side of a sheet of paper write the attributes that might be important to your customers. These might be speed, low cost, quality, value, geography, company longevity, innovation, expertise, people, specialties, or any number of other attributes. For a quality list, ask some key customers what attributes are important to them. At the top of the sheet, write a list of competitors. Include your company in the list (see Figure 6.1).

Work through the list, one attribute at a time. Number each box in the positioning matrix with a one to indicate the leadership position, a two to indicate the second position, a three for third, and so on, working through each competitor and yourself. There will be some ties and some unknowns. Be as honest as you can be. The places where you are number one are the leadership positions on which you need to capitalize. Where you are ranked low are the areas that you may need to shore up or at least build a defensible position. Of course, some of these attributes are more important to the consumer than others. If the attributes are strategically vital, you need to work on them aggressively. If they are optional, you can focus your energies on more important issues.

You'll find when you do this exercise that you will be tied for number one or are number two in certain categories. This is a good time to determine if you can attain sole possession of the number one position. You'll need to ask yourself what it will take to gain number one status, what it will cost, and if it will be worth it. You may also find that no single organization holds the number one position. We have worked with companies that made this determination. In some cases, they claimed the number one position based on some legitimate criteria, and then went about making it true through their actions.

If you have elected to do the positioning process internally, you should verify your results with quantitative research if you can afford to do so. Quantitative research is usually the only way to get a completely objective point of view.

A position that works for you is one that works for your markets. Holding a unique position is only valuable if your market cares about

	ACME ELECTRIC	JONES ELECTRIC	B&E ELECTRIC	SMITH & FRANKS, INC	OUR FIRM
LOW COST	1	2	3	2	3
HIGH SERVICE	3	3	2	2	1
SPEED	3	3	1	?	2
GOOD CREDIT TERMS	?	1	?	2	1
AVAILABILITY	3	?	1	2	3
LOWEST MARK-UP	1	?	3	?	2
CLEAN-UP	?	2	3	3	1
LONGEVITY	3	2	1	3	2

Figure 6.1.

the position. That is what we described as "relevance" in chapter 4. The market must care about the position enough to consider it worthwhile to do business with your company. Thus, within any given market, some positions are more powerful than others. It may be that a company must dominate in more than one secondary position to offset the power of a competitor holding the number one position in a category deemed very important to the target market.

A good position is

- important to the customer,
- not easily duplicated,
- defensible,
- easily communicated,
- easily remembered, and
- affordable or accessible to the target market.

Some of the more obvious positioning strategies—or attributes—revolve around price, speed of service, friendliness, professionalism, ease of transaction, ease of use, and location. Others that can be considered include user segmentation (e.g., this product is designed for busy, working mothers), place of origin (made in the United States or locally owned and operated), usage patterns ("the breakfast drink" or "nighttime relief"), innovative products (Apple), or even product packaging ("the product in the tall, brown bottle" or "packed in totally recyclable packaging").

A position that will become more and more important over time will be the sustainability of your company, your products, and your packaging. Because of increasing global climate change, the declining availability and rising price of oil, and the poor environmental standards of some overseas competitors, sustainability will grow increasingly important as a positioning attribute. On a sustainability measure, locally made products and locally owned stores will become more valuable because they keep money in the community, support more jobs, and reduce shipping, which in turn reduces the cost to the environment. Sustainability is more than just environmental stewardship. It also includes taking care of your employees by paying a living wage, providing health care, and allowing participation in the governance of the company. Of course, if your company is sustainable, it's also making a profit. Information about things your company can do to build a position of sustainability can be found many places on the Internet.

When you have completed the positioning matrix process, you should be able to put together a positioning statement. Usually a positioning statement for a company states the leadership position that the company holds relative to its competitors. Yours might be something like the following: "[Your company] is the leader in providing [key attribute or attributes] [products or services] to [customers] in [region]." So, it might read, "The Jones Company is the leader in providing fast delivery of brand-name pharmaceuticals to hospitals in the Northeast U.S."

The final positioning statement is a key point that you want to express through all company touch points as part of the brand foundation.

Positioning Exercise 2: Unique Brand Components Process

Determining your unique brand components (UBCs) is another way to develop positioning. Looking at differentiators that come from inside, rather than from the competitive perspective, as you do in the positioning matrix process, can help you position on unique internal differentiators that separate you from the competition because they are unique to you. In other words, the UBC process is one way to get to both personality and position traits that might not be obvious through the positioning matrix and the personality development process described earlier. It is also a good way to develop creative concepts for advertising.

The best method we know to determine what your UBCs are is to look at them as facts or true statements about your organization or product. We suggest looking at the history, people, stories, values, and the products/services of your company separately to determine what differentiators exist in those categories.

Here's How

In this process you gather a group of people from your organization to take part in a controlled brainstorming session to determine these differentiating truths of your company. These people will form your "brand team." The brand team participants in this exercise should be drawn from throughout the organization, with an emphasis on having people that interact with the outside world. In other words, choose people who are "touch points" in your company. These may be people that interact with customers or vendors such as the head of sales, the receptionist, and the president of the firm. The exercises that they will go through will take a half day and should be done offsite and without phone interruption to ensure that the participants are focused on the project. The group should be no larger than a dozen people. You will also need a facilitator and a notetaker with a large pad.

In this exercise, first have the brand team list all the things that are true about your company in the following five categories:

- History
- People
- Stories
- Values
- Products/Services

You should generate a fairly long list. Be sure that the highest levels of management always talk last. This will prevent employees from just echoing what the boss says. Also, be sure that there are no arguments. If someone believes something to be true in that specific category, then it goes up on the note pad. If someone thinks the opposite, it goes up, too. Make this a fast-paced exercise. Try to finish each category in 15 minutes. Post the sheets on the wall as the note taker fills them.

Under "history," you might initially have such mundane-seeming facts as "four locations in 10 years" or "started with an idea written on a napkin." Or you might have that you are the oldest, or newest, company in your field in your market area. Under "people," you may have something like "our founder was an industry pioneer" or "all of our employees volunteer time in the community." For "stories" you might have something like "our founder fell from a ladder and broke both legs . . . laying in bed recuperating he figured out how to build the business." In shorthand for the notetaker it might say, "ladder story" (see Figure 6.2). In any case, you should be able to develop dozens of items under each category unless you are a brand new company. Try to stay organized, but don't get too hung up on what category things go under. It doesn't matter because you'll be looking at all your lists together at the next step.

After you have gone through the five categories, start reducing the size of your list. Go to each truth about your company and ask if it's relevant to your market or if it differentiates you from competitors. If it is relevant or differentiating, leave it up there. If not, strike it and move on. People may disagree as to relevance or differentiation. If there is disagreement, leave it up and move on to the next one.

After you have gone through all five categories, start over on trying to reduce the size of the list. Your objective will be to narrow the list down to five or fewer UBCs. Five UBCs are plenty to work on. When there is disagreement in this second round, open up each point to a discussion and go with the consensus. You may have to go through one additional round to get to your five or fewer. Let the participants know that they may have to give up something that is important to them to get to five (see Figure 6.3). Remember, check these to make sure that they are relevant to your customers.

When you are down to five, the final UBCs may end up being all from one or two categories, and the other categories may end up with none. You don't have to have one from each. These five unique brand components can become the basis of your marketing communications based on differentiation from your competitors or they can help you uncover positioning attributes that you may have missed on your positioning matrix.

When you have finished narrowing your list down to five or so, take them one at a time and ask the team, "If these things are true, how can we make them more true through marketing communications, product or service design, or operations?" Compile a list for each UBC in each of these three areas. If time is short, you can work on the brainstorming of making your truths more true in another session with the entire team or a subset of it. These lists will then need to be prioritized later by the brand team to become action items in your brand building process.

Defining Your Brand Promise

Just as there are scores of definitions for branding, there are many for brand promise. For the purpose of building your brand foundation we think that the following explanation works the best:

A brand promise doesn't address *what* you do. It expresses *how* you do it.

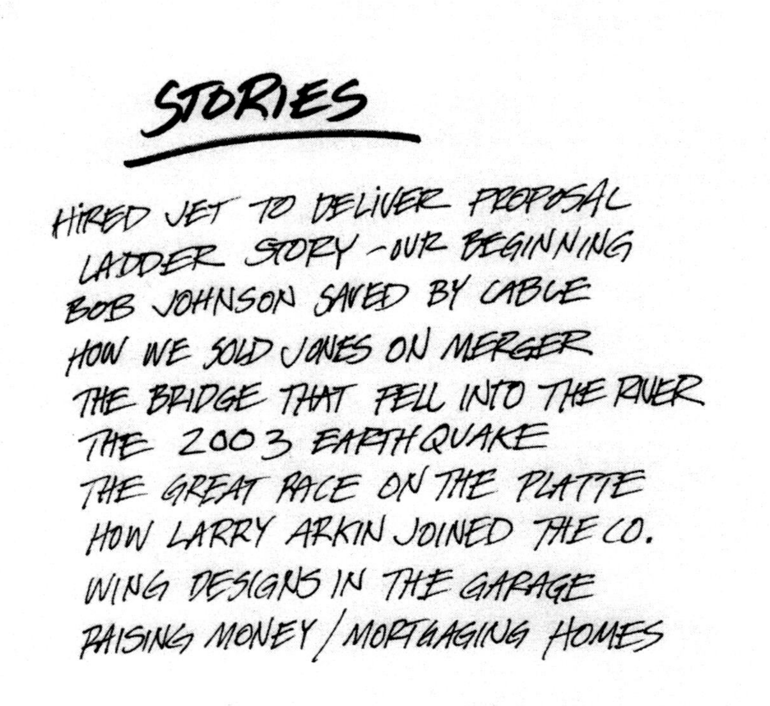

STORIES

HIRED JET TO DELIVER PROPOSAL
LADDER STORY — OUR BEGINNING
BOB JOHNSON SAVED BY CABLE
HOW WE SOLD JONES ON MERGER
THE BRIDGE THAT FELL INTO THE RIVER
THE 2003 EARTHQUAKE
THE GREAT RACE ON THE PLATTE
HOW LARRY ARKIN JOINED THE CO.
WING DESIGNS IN THE GARAGE
RAISING MONEY / MORTGAGING HOMES

Figure 6.2.

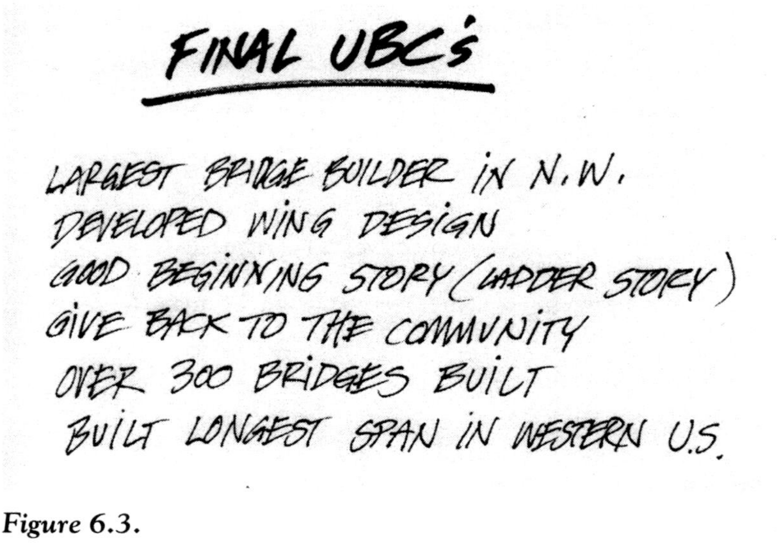

FINAL UBC's

LARGEST BRIDGE BUILDER IN N.W.
DEVELOPED WING DESIGN
GOOD BEGINNING STORY (LADDER STORY)
GIVE BACK TO THE COMMUNITY
OVER 300 BRIDGES BUILT
BUILT LONGEST SPAN IN WESTERN U.S.

Figure 6.3.

Remember the story of the two law offices in chapter 3? Both firms created expectations during the initial brand experiences by how they went about being lawyers. The expectations created by Firm B were the result of defining what you wanted people to expect and how you would meet those expectations. Firm B could charge more because they looked and felt like they were *worth* more. In other words, they had a higher perceived value because each touch point reinforced the preceding one and the following one. Each touch point not only reflected the personality and position of the company but also fulfilled the implied or direct promise that the law firm would take care of you every step of the way. In other words, you could trust them to consistently take care of the details.

Here's How

At this point you want to consciously determine what expectations you wish to create with your customers. What expectations do you want your customers to have? What promise can you make to your customers to effectively reflect your company's values, position, vision, and mission? What is it that you will try to do consistently and reliably every time? Easy possibilities include "always on time delivery," or "never out of stock," or "100% no-questions-asked guarantee," or "we'll treat you with respect every time," or "we'll leave you car cleaner than when we got it." There are endless possibilities, some simple, some complex and subtle.

A company promise isn't always something that you trumpet to the world, although it can be. In fact, a consistently and quietly delivered promise has great potential to build a powerful word-of-mouth marketing buzz, as we demonstrated with the anecdote about Dutch Bros. Coffee. Because promises often reflect your most deeply held values, they aren't necessarily something you brag about. If values are deeply embedded into a company, they are quite evident to customers and employees alike without being mentioned. For instance, if your promise is to always treat people with respect, it will be evident not only in the way that you treat your customers but also in the way you treat your vendors, employees, and competitors.

You will need to determine what promise you can make that your customers will care about. Like we said earlier, the promise can be an internal statement that reflects your values and might never be heard by outsiders. You can have a brainstorming session with your employees to focus on your values, your customer expectations, and your promise to deliver the values. Or you may uncover a great promise when you explore your values in the UBC exercise.

Often companies or industry associations do research to determine what is important to customers. Additionally, the positioning exercises mentioned previously will help you to define your promise. Once you determine what the promise is, you need to write it down and train everyone in your organization including yourself to live it. Always. If you slip up for some reason, you need to do more than make it up to your customers. Promises should not be broken.

Every employee in your business is responsible for understanding and keeping the promise. If your brand promise is to "always be respectful," you need to make sure that everyone in your organization understands what constitutes "respectful." What actions are they expected to do to show respectfulness? Remember, not every employee will have the same definition. It is your responsibility to spell out what you expect. It is essential that they understand the importance of this aspect of the brand experience. In our view, keeping the brand promise should be a job requirement. Failure to do so could be grounds for dismissal. It is that critical.

Let's say you operate your local transit organization and your brand promise is to "always offer the most friendly bus service." Because, as mentioned earlier, being friendly from a bus driver's perspective can mean everything from actively engaging passengers in conversation and helping old people up the steps to not scowling at them, there can be a wide range of actions that a bus driver might consider to be friendly. If that is the case, then part of the company training program should establish systemwide standards for what it meant, in terms of specific actions, to be friendly.

Important note: Before you move on to the actions outlined in the rest of this book, you should have written down concise descriptions of your personality, position, and promise. If you have the resources, you can determine through formal field research if these are appropriate for your organization and of importance to your customers.

A concise version of a brand foundation is useful so that your employees can easily understand it. If you were a landscaping/grounds keeping company, your brand foundation might look like this in its most simple form:

- *Personality.* Reliable, honest, friendly, easygoing, knowledgeable, creative
- *Position.* The leaders in innovative, personalized landscape maintenance
- *Promise.* We'll add pleasure to your everyday life

Like we said, the list above looks simple. However, when you consider the words that describe your organizational personality, position, and promise, you will want to define what specific actions your employees need to take to make those words ring true. What does it mean to be reliable? How do you demonstrate friendliness? Again, remember the receptionist at Firm B, who engaged the customer in real conversation, with real interest and a genuine smile. Those are all things that could be trained or built into hiring guidelines.

Consider the ambitious promise of the landscaper. It says what you'll do for your customer and how you'll enhance their lives. With a promise like that, you build a pleasant expectation with your customers and you demonstrate to your employees what you expect from them in their treatment of your customers.

CHAPTER 7

Brand Distribution

Now that you have determined your personality, position, and promise, the next step is to imbed these attributes into the fabric of your organization. In other words, you need to "distribute" these attributes through every touch point, or point of contact, that the public has with your company or your product. You can break down the brand distribution function into three channels. These are the marketing communications, product/service, and operations shown on the brand map. Each of these channels is like a branch on your brand tree. Each branch has smaller ones that grow from it. Before we go into detail about each of the three areas of distribution, let's look at what each encompasses.

Each of the three areas outlined below are where you can take action to make your brand excel. The key is to remember that all actions you take as a company will produce reactions. **The *actions* you take should be based on the *reactions* you want.**

Ask yourself how you want people to feel about you. What steps can you take to ensure that they feel the way you want them to? How can you embed your brand into everything that you do as an organization?

Marketing Communications

Marketing communications represent a broad range of overt actions designed to connect with your markets. Public relations, advertising, brochures, and the like are the communications vehicles that most people think about and understand, but they are just a few of the many available options in marketing communications. In subsequent chapters, we will examine these and some of the more subtle communications devices available and show you how to utilize them to communicate your brand.

All distribution points for your brand are important. You can do the best advertising in the world and attract hundreds of calls to your business. But if your receptionist is rude, the wait on hold too long, or the series of phone button-pushing steps to reach your goal too frustrating, your investment in advertising could be money down the drain. You might actually be spending money to hurt your own business.

Products and Services

The nature of your products and services is as important to your brand as marketing communications. Unfortunately, too many companies overlook the branding opportunities presented by products and services. They assume that branding is the province of the marketing department. Yet your products and services are the very basis of your business, so they are essential brand builders. Consistency in products and services are key to building trust. Consistency is what makes brands work.

It is very helpful to think of your products as services and your services as products. If you look at your service as a product, you'll look for ways to standardize the performance of the service so that the process is replicable. Thinking this way allows you to infuse the service with brand attributes and reduces the cost of providing the service. And clients are more comfortable purchasing a service that has worked successfully for other clients.

If you look at a product as a service, you'll look at it differently and perhaps will find ways to make the product as much as an experience as an object. Again, Dutch Bros. Coffee is a good case in point. Dutch Bros. sees coffee as a service they provide to make people's days better. Their baristas are trained to make the act of purchasing of a cup of coffee a pleasant interlude in a busy person's life. Without the experience, Dutch Bros. would be just another coffee outlet.

So go ahead and think about your products and services from the buyer's point of view. The buyer typically goes through an unconscious, multistep process before choosing to buy your product or service. You can break this buying process into four steps. We refer to these as the "four steps to a sale":

1. Awareness
2. Consideration
3. Choice
4. Repeat

Awareness is just that. Before you can create a customer, you need to be on their radar. People need to be aware of your brand. Awareness is where the brand experience begins. Sometimes awareness comes from advertising, through PR efforts, from word-of-mouth, from the media or at tradeshows, or from being on the store shelf. Awareness is your first impression. Because awareness can come from so many different places, it is imperative to make sure that every possible touch point reflects your brand foundation.

Consideration is where the prospect is weighing all of his or her choices. Sometimes this is done quickly and intuitively. Sometimes the consumer goes to great lengths to study the choices available and make comparisons. In this stage, differentiation is critical. Your brand foundation, along with factors like price, convenience, and others, are the basis of your differentiation.

Choice is when the sale is made. Ideally, the customer has chosen you. They are picking your product off the shelf, placing an order, or talking to a salesperson. The customer may still be nervous and unsure of their decision at this point. This part of the experience has to meet the expectations created by the brand foundation as demonstrated through the first two steps. The sale needs to be as painless as possible for the buyer.

Repeat is when a buyer comes back to your product or services and decides to pick you again. This is perhaps the most important and certainly the most often overlooked step in the experience for the customer. The experience must meet or exceed the expectations created by the brand to this point. If not, you have a dissatisfied customer who either returns the product; refuses to pay for the service; or maybe worse yet, badmouths the product, service, or company that provided it. On the other hand, if you meet or exceed expectations, you have a satisfied customer, one who will spread the word and buy from you again. You can create evangelists for your product at this point. Every businessperson knows that a repeat customer is a lot more profitable than a new one— especially one who tells others about your product or service.

In this context, the sale of your product is more than just an isolated, one-time event. It is a complete experience with distinct stages. Your objective, as the manager of a brand, is to ensure that the experience throughout the four steps reflects your brand and that each step leads to the next. Your product will become more like a service, especially at step four, when you have the opportunity to use the product as a vehicle to create a customer that will be a positive endorser of your product or company in the future.

Part of the repeat step is to have a plan for how you will continue to stay in contact with the customer following the sale. Some companies use e-mail, follow up by phone, or make sure the salesperson creates a way to communicate with them regularly. Others put the customer on a mailing list for newsletters or add them to a social network to keep in constant contact. Other organizations make customers aware of presale events and special offers. The nature of your business and customers will play a major part in your decision about how to stay in contact to ensure repeat business.

Operations

Marketing communications is the area that most companies utilize to promote their brand because it's the most obvious and the first thing they think about. But just because it is the most obvious doesn't mean that it is the most powerful. Marketing communications is like gravity. Gravity's force is blatantly obvious all around us, but it can't overpower electromagnetism and pull a magnet off your refrigerator door, because electromagnetism is a stronger, less obvious, force. Operations, as a function within your organization, is like electromagnetism. It is a quiet, almost hidden, but powerful branding force.

Operations encompass a host of business functions. Facilities design and management, human resources, procedures and processes, manufacturing, and fleet management often fall under the heading of operations. In companies where marketing is a department far removed from operations, brand opportunities are easily missed. Operations managers can, and should, have a big impact on the company brand. In the human resources sector of operations, hiring and training are key brand building opportunities. If, for instance, a key personality trait of your brand is "friendly" then it stands to reason that you would look for friendly people as employees. Human resources needs to understand and live the brand so they can influence the culture of the organization and further push the brand through one of the most important touch points: your people.

The same goes for training. If friendliness is a brand attribute, you would break down for new employees what it means to be friendly on the phone and in person. There are many ways to connect with customers in a friendly way. For example, some restaurants, especially in tourist areas, have nametags for employees that say where they are from. Thus, something as simple as a nametag becomes a conversation starter and allows an employee to demonstrate the "friendly" brand attribute through conversations and building a relationship with the customer.

In the following chapters we'll take the three distribution areas—marketing communications, products/services, and operations—and discuss in more depth how you can use them to distribute your brand.

The brand map that accompanies this book shows many of the typical touch points in an organization. Your own organization may have most of these in addition to some specific to your industry. It is a good idea to sit down with a group of people from your organization and brainstorm every single touch point that you have. Then you can use the suggestions in this book to ensure that every single customer experience with your organization is consistent with your brand foundation. We recommend that you codify how you are going to address all the touch points in the brand manual described in the introduction. Your manual can be used as a training tool and a guide for all of your employees. Think of your brand manual as your strategic plan for how you manage your brand.

Here's How: Touch Points for Your Company

Develop a list of touch points for your organization. Think of all the places that a customer or client encounters your organization or brand. Start with obvious things like the Web, advertising, receptionist, salespeople, business cards, signage, and so on, then move on to more obscure ones like your company training manual, your sponsorship of events, employees on boards, and so on. Try to come up with a good list through a brainstorming session. Then ask all the employees to come up with other ideas to add to your list and make it as comprehensive as possible.

CHAPTER 8

Marketing Communications

In the next three chapters we are going to take each of the three main brand distribution channels and discuss the components and subcomponents of each. After most of the subcomponents we will address how to do it yourself with "Here's How" exercises.

One glance at the brand map will show you both the scope of marketing communications and its importance in brand building. As a measure of its complexity, you'll also see that we don't extend every marketing communications initiative to the next level of the map. For instance, look at "advertising" on the brand map. This communication function is not extended to the next level. This isn't a measure of how important we feel advertising is to brand building. Rather, it is a measure of the complexity of the subject. Advertising is, in and of itself, a complex enterprise that can involve highly paid specialists in creativity, media planning, research, design, copywriting, photography, and more. Because it would take volumes to write about all the aspects of advertising, we'll only touch on key points of advertising as they relate to branding. This is also the case with identity, our first topic in the marketing communications area.

Identity

As we said at the beginning of this book, corporate identity is the area most confused with brand. Corporate identity is more than just a logo. Corporate identity encompasses naming, logo, color, typography, design systems (grids), tagline, iconography (design devices, art, photo, and border treatments), and other subtleties. Let's look at each of these parts separately and see how they can influence the brand.

Naming

What's in a name? That which we call a rose
By any other name would smell as sweet.

— William Shakespeare, *Romeo and Juliet* (II, ii, 1–2)

Thus says Juliet to Romeo. Obviously, she wasn't in marketing. If she were a marketer, she'd realize that names are a key part of identity and brand. Large corporations will pay millions to develop and establish a name because they know that a good name is a valuable asset that over time can build equity. It stands to reason that a memorable name makes marketing communications less expensive. Evocative names can also be very powerful in that they can communicate a concept every time they are heard or seen. Descriptive names can tell a consumer instantly what a company or product does.

One of the measures of advertising success is called "top of mind awareness," or TOMA for short. It is a primary objective of advertising to have your company or product as the first one people think of when they are looking for a product or company in your category. A good, easy-to-remember, evocative name can go a long way to creating TOMA.

The marketing business is full of stories of bad naming. Usually these stories are about companies who create a name in one culture or language and then market to another culture or language where the name has a different or unintended meaning. The Chevy Nova is the classic story. "No va" means "no go" in Spanish. Country Mist by Estée Lauder had problems in Germany, where mist means manure, which is generally not a scent that most people consider beguiling.

Company name changes are not uncommon. Over 100,000 names are filed annually in the United States and many of them are not brand new companies. Often, name changes are prompted by mergers or changes in business direction. Sometimes they are prompted by pronunciation difficulty, the expansion to new markets or countries, or confusion between company and product names.

There are many processes to develop a name. Some companies brainstorm using categories of name types. Because there are so many names in use, finding one that is available is difficult. Often, a company will develop thousands of names to get 8 or 10 that are appropriate and available. Once a name is developed, you need to go to the USPTO Web site or have a law firm check its availability. Remember, generally speaking, you can have the same name as another company if the name is not registered in your category. For example, if you have a local insurance company in New Jersey, you can name it Eastern Insurance even if there is another company called Eastern Airlines. You'd have a fight on your hands, however, if you named your firm the Coca-Cola Insurance Company.

There are five common name types:

- *Proper nouns* are usually names of people or places, such as Gibraltar Savings, Macy's, and Smith & Wesson.
- *Descriptive names* are the names, like Speedy Print or A-1 Auto Glass, that describe the nature of the company.
- *Acronyms* are very common. Examples are IBM, AT&T, and the well-known numerical/alphabetical version, 3M.
- *Metaphorical* names include Thunderbird, Nike, and Target.
- *Invented* names, or neologisms, a category that is getting more common as we run out of nouns and verbs, include Xerox, Cialis, and Intuigy.

There are also combination names that combine aspects of two name types. Sensormatic, TecLabs, and Microsoft are examples of composite word names.

It is likely that many of you, as you read this, are already in business and already have a name for your company. If you feel that your current company name isn't the best name, does it make sense to change it? Whatever your current name, it probably has some equity value, especially if you have operated under the name for a long period of time. The question you need to answer is whether the current name equity value outweighs the potential equity value of a name change to a more memorable or easier to pronounce name. This is a difficult and often expensive decision and we recommend that it should be made with advice from a professional.

Other questions to ask before embarking on a name change are, Does your name no longer reflect what you do? Is your name so difficult to say that your own staff answers the phone with just part of the name or a shortened version of it? These are warning signs that maybe a name change should be considered.

You can examine the name change question and inform your decision about maintaining or changing your company name by undertaking name equity research. Conduct this research in your marketplace to determine how well your name is recognized and what values and attitudes are connected to your company name. If your name is well recognized and has strongly attached positive values, then leave well enough alone unless other factors, such as a change in business direction, overshadow the positive equity.

One last thing about company name: Whatever the name, you should work hard to attach key brand value words to the name (note: go back to the "personality" words in your brand foundation for a good place to start). Your objective is to connect your name to these key brand values. Ideally, when your market hears your company name, they should immediately, unconsciously or not, associate those words with your company.

Here's How

Naming, like any other component of identity, should be developed with specific criteria that reflect the brand foundation. Criteria used for naming includes memorability, noncorruptibility, pronouncibility, evocativeness, differentiation, and descriptive attributes. The criteria used should be the basis of judging the adequacy of the name, or, it sometimes is used to help frame the naming process.

Logo

The logo is the most obvious of the identity system components. Some logos are so ubiquitous that even small children recognize them. Companies value their logos as important company assets and will spend hundreds of thousands of dollars to create and defend them.

In the United States alone, more than 800 logos are registered daily with the U.S. Patent and Trade Office (USPTO). With the high number of corporate logos registered both formally with the USPTO and informally registered through first use, or common law use rights, it may seem impossible to invent new ones. Fortunately, registration is by business category and there are 45 international categories. The rule of thumb used to determine if a logo encroaches on another's territory is simple: "Will the public be confused by the similarity?" In that context, if you own a coal mine, and your logo is similar to that of a toy manufacturer or a dog-grooming company, it is likely to be ruled that your logo would be confused with theirs. Nevertheless, one cannot copy a logo directly or too closely.

Because of the limitations described previously, along with the typical physical use requirements of a logo (usable on both a business card and on the side of a building, for instance), many graphic designers consider logo design to be the most challenging of design projects. Some of the logos of international companies cost hundreds of thousands, even millions of dollars to develop. Much of this expense goes to vetting logos to see if they are usable on the international market, ensuring that they are not offensive to certain nationalities or religions, researching the symbolic associations of the mark, and the generation of usage standards.

The ideal logo symbolizes the essence of the brand. Sometimes, usually in the case of a redesign or a brand new company, the logo is designed to reflect the brand. Other times, as in the case of a company that has an existing logo, the logo comes to represent the brand over time. Thus, the logo actually receives meaning from the brand and comes to symbolize the brand foundation.

Logos can become important company assets with tangible value. In certain cases, the relatively modest cost of a logo design can bring a tremendous return on investment. Depending on the nature of your business, you can spend a few hundred up to several hundreds of thousands of dollars for a logo and graphic standards development. For instance, if you are developing a company that will have a widely viewed, ubiquitous, and memorable logo with constant repetition before the same audience, that logo will continue to grow in value. If and when your company is a candidate for acquisition, the ubiquity and memorability of that identifier will represent a considerable value and may, in fact, represent the majority of the value of your organization.

The decision to keep an existing logo or make a logo change is one that companies bring us frequently. The idea of changing an existing logo often comes up when companies are merging, in the case of a name change or when executives feel that their logo just doesn't reflect their brand. We always tell companies that logo change should be undertaken with caution. Any logo, no matter how bad the design, has a certain amount of equity. The value of the equity must be considered. Ask yourself if a new logo would be more valuable because it is more memorable or if the change would be too costly to be worth it.

There are hidden costs associated with logo change and they can be enormous. All printed materials must change. Your building and fleet signage need to be altered. Introduction of the logo to everyone from vendors to the public can be costly. Increased advertising expenses to promote the new look can add up. If, all things considered, you feel it is time for a change, there are several things to consider about symbol development that are brand related. Many books have been written about logos and logo development, so we won't go into the intricacies of corporate identity design. Instead, we'll tell you about two things that you should consider when judging a logo symbol or a logotype:

Silhouette

When we talk about the silhouette of a logo we are referring to the shape of the symbol. Silhouette is an important consideration in what the mark expresses. Rounded marks appear softer, more passive and less aggressive. Conversely, marks with sharp edges appear aggressive, bold, young or active. Marks with vertical and horizontal stability appear grounded. Marks at an angle or tilted appear more bold and assertive and often younger. A designer can also imbue a mark with a touch of humanity by affecting how the edges appear. Soft, wavy, or rough edges can give a handmade immediacy to a mark. Pyramidal marks seem strong. Ovals, especially tilted ones, have come to symbolize modernity.

Figure 8.1. Round (Open Adoption)

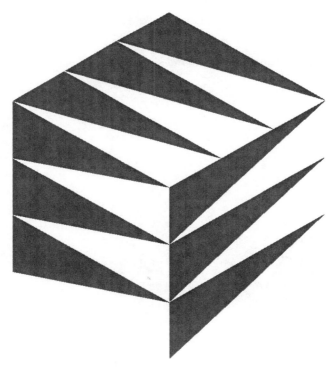

Figure 8.2. Sharp (POSdata)

Figure 8.3. Vertical and horizontal stability (Lane Mammography)

Figure 8.4. Angled or tilted (National Warranty)

Figure 8.5. Soft, wavy, or rough edge (Eugene Parks)

Figure 8.6. Pyramid (States)

Figure 8.7. Tilted oval (Peterson)

All logos by Funk/Levis & Associates, Inc.

Gestalt

Many designers believe that logos have a gestalt of their own. Gestalt, in logo design, means that the whole is greater than the sum of the parts and also refers to the energy of the mark. There are two types of gestalt in logos. There is inner gestalt, where the parts seem to converge on the center. Outer gestalt is described as the situation where the logo seems to emanate or burst from the center. This is a very active and young stance for a logo, while inner gestalt tends toward a more sedate and centered feel.

Figure 8.8. Inner gestalt (OAN)

Figure 8.9. Outer gestalt (Downtown Eugene)

All logos by Funk/Levis & Associates, Inc.

Logo design is best undertaken by outside, professional designers with written criteria in order to judge the submitted designs against objective, rather than subjective standards. These criteria should be put down in writing (i.e., general and specific criteria) for the design. Specific criteria is that criteria that is specific to your company and should include words from your brand foundation. For general criteria, we refer to the time-tested criteria from Yasaburo Kuyama, author of *Trademarks & Symbols* (Van Nostrand Reinhold, 1973). In the book, Mr. Kuyuma states there are nine general criteria for logo design:

1. *Suitability of content.* Does the mark make sense for the company and brand?
2. *Suitability to the media.* The mark should work for all applications and types of media (i.e., Web, print, neon, electronic, etc.).
3. *Distinctiveness.* Is the mark clearly different from others in the same field?
4. *Contemporaniety.* Will the mark endure and still look fresh in 5 or 10 years?
5. *Memorability.* Is the mark easy to remember at one glance?
6. *Reliability.* Does it suggest that you'll keep your brand promise?
7. *Utility.* What happens if the mark is turned upside down or backward? Will that have an undesirable effect? Is it difficult to determine the top? Does it animate?
8. *Regionality.* Does it reflect your nationality or region? Is this good or bad?
9. *Color individuality.* Is the color distinct from other firms in your field? Is it a color that reflects your brand? (We discuss color more in the next section.)

Judging logo design using preestablished, specific, and general criteria removes much of the subjective opinions and makes the process much easier. It is also initially easier to judge comparative logo design options in all the same color, preferably black. This will remove color subjectivity from the evaluation process and allow evaluators to concentrate on design. By the same token, it is often better to first judge symbolic marks without accompanying type.

Color

Color is strongly linked to brands. Throughout the world, everyone knows that Coca-Cola is red and white, Kodak is chrome yellow, IBM is blue, and United Parcel Service owns brown. Typically companies have a primary color. The primary color is often paired with a secondary color. These identity colors are often supported by a palette of colors that are reflected in everything from interiors of facilities to advertising and associated products and services.

Color is a very powerful and memorable force in branding. Here's an instructive example from a personal incident of one of the authors. This incident describes both the power of color and the power of the brand experience of ice cream. I suppose it also describes the connection between color and brand and maybe the fact that a child was fed too much ice cream in his youth. When my son was barely old enough to talk, I gave him some crayons to play with. He was scribbling on a piece of scrap paper with the crayons, grabbing one and then another like kids do, when he happened to grab pink and brown and scribble a few strokes with each in quick succession. "Look, Dad!" he shouted. "That's nice," I said. "What is it?" He looked at me like I was a total idiot and firmly explained, "It's the Baskin-Robbins ice cream store, you silly."

Identity designers are like chefs in that they can combine symbolic ingredients to create different recipes. The ingredients that the designer works with include color, typography, and the symbol itself. Like silhouette, color is expressive and carries meaning. The meaning of different colors varies by culture and changes cyclically. Colors gain and lose popularity, although blue remains the world's most popular color, and is far and away the color picked by men as their favorite. Color preference is determined by a host of factors including eye color, age, gender, education, and culture. Blue-eyed people tend to prefer cool colors while brown-eyed people prefer warm ones. In certain countries green is a symbol of richness; in others it is considered cheap.

Is America truly red, white, and blue? Not necessarily! Studies such as the Global Market Bias Research by Cheskin, MSI-ITM, and CMCD Visual Symbols Library show that "people associate color with emotions, products, companies and even countries." Their study in 2004 showed that a healthy percentage of many nationalities—most notably the French, Brazilians, Russians, and Germans—see America as black, a color associated with anger, aggression, and coal. The authors suggest that America would be well served to borrow some goodwill from the color yellow, a color that suggests sun, happiness, and warmth.

The same study showed that people throughout the world thought of BMW and Mercedes as black; Coca-Cola and Marlboro as red; McDonald's as yellow and red; MasterCard and Kodak as yellow and orange; and Pepsi, Levi's, Microsoft, and IBM as blue.

The physiological aspects of color are well known and utilized by direct marketers. When surrounded by too much of one specific color, people will unconsciously seek out its complement color. Therefore, in the spring in the Midwest, when farmers are surrounded with growing green things, sending direct mail with a healthy dose of red is more effective than sending out green mailers. You can test this color complement phenomenon yourself by painting or coloring a quarter-sized circle of color in the middle of a white sheet and staring at it under good light. After a minute, take the sheet away and replace it with a blank white sheet of paper. What you'll see is the exact opposite color on the color wheel.

In logo design, you'll often find that one specific criterion seems incompatible with another one. This is where we come back to the analogy of a good designer being like a chef. The mixing of ingredients or criteria becomes important. Let's say that you have a company that expresses its personality as aggressive and professional. Your designer has come up with a design for the symbol part that has a sharp, bold silhouette. That handles the aggressive part but seems to need some tempering to give it a professional aspect. You can match a traditional, Roman-based typeface to add gravitas. You can also choose a non-aggressive color, such as blue, to further tone down the aggressiveness of the overall identity.

You should make sure that your logo designer is cognizant of the cultural and physiological connotations of colors and symbols, especially if you ever intend to market on an international basis. An excellent reference on color is the *Symbol Sourcebook* by Henry Dreyfuss, published by McGraw-Hill.

To demonstrate how color works with a brand foundation, let's look at a fictional wholesale nursery that specializes in growing pear trees. They have a brand foundation that looks like this:

- *Personality.* Reliable, honest, friendly, easygoing, knowledgeable, creative.
- *Position.* The leader in organically grown pear species.
- *Promise.* We know pears like no one else.

What are the colors that show "reliable, honest, friendly, easygoing, knowledgeable, and creative"? Friendly, easygoing, and creative are best represented by warm colors such as shades of red, orange, or yellow. These colors are also conveniently the general colors of many pears. Honesty and reliability are blue. Knowledgeable is represented by green. This could indicate that our fictional nursery should have a two- or three-color palette of warm colors accented by a cool green/blue, assuming that the dominant personality trait is the friendly, easygoing nature of the company.

To gain consistency of color, we recommend that you make sure to write down your company's primary, secondary, and tertiary colors by PMS (Pantone Matching System) number, RGB mixture percentages (red, green, and blue, used in Web development), and by CMYK mixture percentages (the four-color process colors: cyan, magenta, yellow, and black). This might also apply to vinyl colors, composite laminate colors, paint colors, and other color systems. Codifying these and other design elements in your brand manual will help ensure consistent visual application of your brand. We are constantly amazed at the inconsistency in color application. Test this for yourself: gather several business cards from different people in your organization and several letterhead samples. Then open up your Web site and look at your sign. Do the colors seem consistent? Are all the business cards the same? Would your customer base be able to identify your company colors?

Grid

Ever heard of a grid in reference to design? No? There's a reason. In a finished design, it's invisible. But that doesn't mean it's unimportant.

Humans are very perceptive. They notice details that don't even register on the conscious level, but still act to influence feelings and opinions. That is why attention to every detail is critical to successful brand building. One detail that is often overlooked is the underlying structure of your visual marketing tools, what designers call the grid. The grid is almost never visible in final marketing materials. It usually is only visible on the designer's computer while the piece is being designed. The grid determines placement of columns of type, headlines, and pictures. The objective is to have a consistent presentation of your brand in every visual manifestation (see Figures 8.10 and 8.11 for examples of a hidden grid and its application). Most magazines are designed with a three-column grid. Most newspapers have five or six columns. Grids can be simple or complex, flexible or rigid. The key to designing a grid for your organization's materials is to design it to be a differentiator from the competition and reflective of your brand. Thus, if your brand is modern and forward thinking, your grid should probably not be a simple three-column grid, but something a bit more modern feeling and not so traditional.

Here's How

Work with a professional designer to develop a grid that reflects your brand platform to be used for your printed marketing materials. The grid then should become the basis for your printed materials and will give you a consistent look and feel throughout. Once you settle on one, stick with the grid system in work produced internally and externally. It's important to have your grid spelled out in your graphic standards document.

Typography

There are thousands of typefaces available today, with hundreds more created annually. The simplest classification breaks down type into three categories: serif, sans serif, and display. These classifications can be broken down further into old style, traditional, and modern. Serif typefaces, those with "feet," have been around since before the Roman Empire. Sans serif fonts (without "feet") have been used since the early nineteenth century. In general, fonts with serifs communicate tradition, whereas sans serif fonts communicate modernity. Examples of widely used serif type families include Times Roman, Garamond, and Bodoni. Popular sans serif fonts are Futura, Helvetica, and Franklin Gothic. In contrast, Display faces include any number of variations of calligraphy; of fonts derived from technological numeric displays and themed faces, such as cowboy/western-inspired fonts.

Type styles function like silhouettes and color in that type also has a symbolic function. Sans serif fonts can suggest contemporary ideas. Old style fonts can symbolize stability. Display faces can have any number of connotations. In addition, designers may create their own fonts for a specific identity situation.

When designing corporate identities it is possible that the unique font becomes the primary identity (FedEx, Coca-Cola, IBM) and in other cases the symbol becomes the primary identity (Apple, Nike, Mercedes-Benz). Consider your choice of type carefully. When designing corporate identities with symbols, use caution to avoid typefaces that call too much attention to themselves and thereby direct attention away from the symbol or compete graphically with it.

Here's How

You should create written graphic standards for your brand. These standards should include uses of typography, color, and the logo. When thinking about typography, consider the following: Do you use all capital letters in your headlines, or is just the first letter capitalized? How do you treat captions? Subheads? Headlines? Body text? Do you center type or have it flush left? What typefaces best represent your brand personality?

Tagline

Look at any consumer magazine and you'll see that almost every advertiser has a tagline near their logo. The company tagline is the verbal identity. It is the last thought that you want to leave with the viewer. It sums up your campaign, your ethos, and your brand. Taglines are the more flexible part of your company identity. Unlike logos, which evolve slowly,

The Publication Grid

Gutters
Spaces between columns are called gutters. They help to break up long columns of text, and can create rhythm on a page layout.

Columns
Vertical divisions of space, where text, photos, or white space can occur are called columns. The more columns on a page, the more complex page layouts can be.

Horizons
Horizontal divisions of space, that help lead the eye of the reader across a page, help break up long columns, and provide places for photos to crop. They also create modules that allow for smaller divisions of space.

Margins
Margins are the outer edges of a page, where no "live" matter should print. In many publications, margins are necessary to accommodate bindery limitations.

Module
Modules are the smallest units of space within the publication grid. They often contain captions, spot photos or small illustrations.

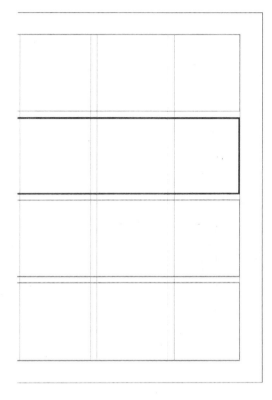

Figure 5.10

Application of the Grid

Headlines
Headlines define the topic of a page, and focus the reader's attention on the primary subject matter.

Spot Photos
Small photos or illustrations may span one or more modules. In this case, the photo breaks the grid slightly, to create tension, while still adhering to the overall page structure.

Body text
The four-column grid is used to contain the main text of this article. The type size and line length have been adjusted to maximize legibility.

Poplars

Lorem ipsum dolor sit amet, consectetuer adipiscing elit, sed diam nonummy nibh euismod tincidunt ut laoreet dolore magna aliquam erat volutpat. Ut wisi enim ad minim veniam, quis nostrud exerci tation ullamcorper suscipit lobortis nisl ut aliquip ex ea commodo consequat.

Duis autem vel eum iriure dolor in hendrerit in vulputate velit esse molestie consequat, vel illum dolore eu feugiat nulla facilisis at vero eros et accumsan et iusto odio dignissim qui blandit praesent luptatum

zzril delenit augue duis dolore te feugait nulla facilisi. Lorem ipsum dolor sit amet, consectetuer adipiscing elit, sed diam nonummy nibh euismod tincidunt ut laoreet dolore magna aliquam erat volutpat.
Ut wisi enim ad minim veniam, quis nostrud exerci tation ullamcorper suscipit lobortis nisl ut aliquip ex ea commodo consequat. Duis autem vel eum riture dolor in hendrerit in vulputate velit esse molestie consequat, vel illum dolore eu feugiat nulla facilisis at vero eros et accumsan et iusto odio

dignissim qui blandit praesent luptatum zzril delenit augue duis dolore te feugait nulla facilisi.
Nam liber tempor cum soluta nobis eleifend option congue nihil imperdiet doming id quod mazim placerat facer possim assum. Lorem ipsum dolor sit amet, consectetuer adipiscing elit, sed diam nonummy nibh euismod tincidunt ut laoreet dolore magna aliquam erat volutpat. Ut wisi enim ad minim veniam, quis nostrud exerci tation ullamcorper suscipit lobortis nisl ut aliquip ex ea commodo consequat.

Poplars are a category of trees which include aspens, cottonwood, and sycamores.

Poplars are deciduous, and often grow near rivers, lakes and other sources of water.

Captions

Captions are often set in italics, with extra space between lines of text, to add contrast to the page. This helps differentiate them from body text.

Figure 8.11

taglines have shorter life spans in most cases. Sometimes taglines are related to a specific marketing campaign; sometimes they are closely tied to the brand itself and remain a part of the logo block for decades. In general, consumer brands aimed at younger audiences change their taglines more often than traditional companies marketing to older demographics.

There are different styles of taglines. Some are imperative, like Nike's "Just do it." Some are descriptive, like General Electric's "We bring good things to life," or Seiko's slogan for its Pulsar watches: "Where substance meets style." Some are interrogative or challenging, like Capital One's tagline, "What's in your wallet?" Some last and last and become part of the cultural vernacular, such as "A diamond is forever."

Whatever style tagline you use should be consistent with your brand. The bold, brash, in-your-face Nike tagline perfectly reflects their aggressive, edgy brand. Renaissance Hotels & Resorts builds their brand around the differences in their hotels, rather than the McDonald's-like sameness and consistency message. Their tagline and the basis of their advertising is "Uniquely Renaissance." Another company, Kyocera Copiers, position themselves as a friendly, low cost leader. They use the stamp "People friendly" in their ads. They also use the tagline "The new value frontier."

We recently developed a tagline for a company that developed software and hardware applications for lumber mills. Well, truthfully, we borrowed the tagline from one of their archived ads and realized that it reflected the brand perfectly. The client wanted to differentiate their software and hardware from similar-seeming systems that used the same basic technology. They went through training to change the way their salespeople sold products. Their objective was to turn a competitive, price-based buying process into a consultative process. They taught their salespeople to ask questions and find the issues that mattered to the potential buyers. Only then, when they fully understood the customer's "pain," did they offer a solution. Sometimes they realized that solution was not their product, and they told the customer so.

We suggested that they use their old headline, "First, we listen," as their tagline. To establish meaning for the brand concept behind the tagline, we developed our representative icon: a serious, ruggedly handsome man in a company hardhat, listening intently as a customer tours him through a lumber mill. Then we built a series of ads and brochures explaining our brand differentiation of listening rather than selling. The tagline thus took on meaning and established them as a different kind of company in a market of perceived product parity.

Here's another example, this time of how you can marry a logo symbol, type, color, and tagline. Our firm developed a logo and tagline for a non-profit organization called ShelterCare. ShelterCare helps homeless and mentally disturbed street people by sheltering and training them. The logo shows three dancing forms (happy people) in the outline of a friendly, minimalist house (see Figure 8.12). The color is an electric, friendly blue. The typeface is a modified ITC Galliard style, a friendly, warm, serif face. The tagline is "Hope is here." In this case the symbol, color, and type shows friendliness, approachability, and even a bit of fun, modified slightly by the serif face and the blue, which both show the serious nature of their work. The message of the tagline complements the serious aspect of their identity. It says that ShelterCare is a warm, inviting environment that does serious work.

Figure 8.12.

Here's How

Taglines should be as short as possible. They should be developed using written criteria based on your brand platform. What do you want your tagline to say? Do you want it to be redundant with the logo mark, reinforcing the logo, or do you want it to express another part of your brand? Do you want it to reinforce a marketing campaign? Or vice versa? To create your own tagline, generate a series of options and test them against your written criteria. Test them outside your organization to ensure that they don't have unintended meanings or connotations.

Iconography

Iconography has a broad meaning. When we talk about iconography in a corporate identity sense, we use the word to refer to all the graphic elements that represent your unique brand. Your company should have its own iconography reflective of your brand. That style should be codified and used consistently through all marketing communication elements.

Your brand should be reflected in how you use illustration and photography. Determine how you will consistently treat photographs or illustrations. Will you use borders? What kind of border will you use? How will the photos be cropped? Will they have drop shadows to make it appear that they are floating on the page? Will the borders be straight or with wavy curves? What subject matter in the pictures, or what style of illustration will reflect your brand best?

Here's How

Iconographic elements should be codified and included in your brand manual. These elements include advertising borders, use of initial capital letters to start copy blocks, curves, blocks of color, and any other visual element. Each element should reflect your brand foundation. Again, it is helpful to develop your style with written criteria.

Advertising

Advertising is the most commonly thought of marketing communication vehicle. It certainly is the most ubiquitous. Some experts estimate that each individual is exposed to over 3,500 direct advertising messages per day in the form of ads in magazines, radio, newspapers, billboards, bus-boards, TV, and any of the numerous intrusive advertising venues from urinal ads to subway posters. The list goes on: Web advertising, direct mail, telephone solicitation, license plate frames, decals and stickers, logos on pens, shirts, shoes, and so on. We are surrounded by advertising; inundated day and night at home and in public. Because it is so intrusive we've developed defense mechanisms to avoid it. We simply ignore it and filter it out of our lives. At least we do on the conscious level, but who knows how much of it is imprinted on our subconscious?

Although most advertising is ignored or goes in one ear and out the other, companies still feel compelled to advertise. Often they advertise because their competitors are and they don't want to be left behind. But advertising is expensive, so it is imperative that it be done with care and deliberation to get the highest return on investment.

There is far more to consider than the things that we've already covered when it comes to advertising. Advertising is a complex endeavor, and many excellent books have been written on the subject. Good advertising considers who, what, why, when, and where in terms of the audience. It considers media, timing, psychographics and demographics, competition, and much more. It is too expensive to leave your advertising to nonexperts. It is also too expensive to leave advertising to firms that don't understand that good advertising must be consistent with your brand.

In the cacophonous world of advertising, you can stand out if you stick to your brand. Advertising is often the first place where many of your customers—and potential customers—come into contact with your brand. Advertising may be the portal through which people enter the world of your brand and begin their relationship with you. If that's the case, you want a good first impression that is consistent with the brand experience you have developed in your brand foundation.

If you've read this far, you already know many of the things you need to consider to create brand-relevant advertising. Things like identity, grid, color, typography voice, and all their subsets should be integrated into your ad designs. In other words, your ads should feel and sound like your brand and should reflect your brand foundation.

From a branding point of view, not only does the look and content of your ads have to be consistent with your brand, but all aspects of advertising also have to be considered in light of the brand. For instance, choice of media are an important brand consideration. Is the "neighborhood" of the medium, be it social networking, newspaper, radio, television, or magazines, consistent with where a company like yours would be? By neighborhood, we mean the specific medium's environment. For example, a classy health club would be ill advised to advertise in a penny advertiser throwaway. Your brand foundation represents your standards.

Regardless of the benefits of the bargain prices you might get from media that is below your brand standard, where you advertise is as important as what you advertise. If you know your market well, you can make media buying decisions that are appropriate to your position.

Advertising has changed drastically since the advent of the internet and social media. Internet, social media, and e-mail advertising are rapidly growing in terms of sophistication, hypertargeting, and interactivity. Because technology is changing so quickly and technological change brings new advertising opportunities, it is imperative to keep a close eye on the various mediums to ensure that your Web site and Web advertising is consistent with your brand. It is easy to get distracted by technological bells and whistles that may take you off your brand look and feel. Don't do something that deviates from your brand, no matter how cool you think it looks. Your brand equals the sum total of all the various brand messages at all touch points. Don't include any that won't reflect a consistent brand.

Here's How

Every time that you design an ad, check it against your brand foundation and your identity components. Ads should start with a creative brief that describes the intent and objectives, the market needs, the demographics, and psychographics of the reader or viewer, and the essential ad elements. A typical creative brief also describes the medium (e.g. for print: size, paper stock, colors, deadlines, etc.) and the budget. We recommend that, in addition to starting every ad project with a creative brief, you create an ad checklist with which to proof each ad. Much like a pilot's preflight checklist, an ad checklist makes you slow down to ensure that you are sticking to the brand program. You should include things like, Does the ad have an appropriate headline? Do we use our brand words in the copy? Are we using our approved graphic and brand standards in this ad? Are we using our brand colors? Does the ad and copy reinforce our brand? Other questions can be added but these are a few to help you get started.

Collateral

The word "collateral" has come to mean all the printed and specialty sales support material of an organization. Collateral includes flyers, newsletters, brochures, letterhead, business cards, white papers, specialty advertising, and more. All these items should consider the brand and reinforce it. Specialty advertising items such as caps, cups, pens, and so on are easily overlooked as brand reinforcers and are often not purchased with consideration of the brand in mind. Whenever you purchase these giveaways, make sure that they reflect your brand foundation and your other marketing materials.

Companies often run into trouble with their brand in the development of collateral when they have the collateral created without brand foundation guidelines. It is important to make sure that those who are putting together collateral follow the brand guidelines of the company or make a very good case why not to follow it in a particular instance.

Here's How

Every piece of collateral you develop should have a creative brief such as that described in the advertising section. The creative brief should address how the particular piece of collateral reinforces the brand. It should also describe what other pieces of collateral will be used in conjunction with the particular piece.

Public Relations

Public relations (PR) is a critical component of marketing communications, and thus of brand building. PR includes media relations, event planning and management, speech writing, crisis communications strategy, and government relations such as lobbying.

The classic public relations planning model has four steps: research, analysis, communication, and evaluation. Key audiences, messages, strategies, and priorities are determined through research and analysis. Specific communications are designed to implement the strategies. Evaluation methodologies are developed to measure the results.

Public relations professionals start with objectives. What do you want to accomplish with your PR efforts? Do your objectives mesh with your brand? Sometimes you'll have a specific objective that doesn't directly relate to your brand. In those cases, the mantra of the PR person in relation to brands should be, as it is with doctors, "Primum non nocere," or, as we non-Latin speakers say, "First, do no harm." Don't allow your public relations to undermine your brand. Instead, look for ways it can enhance your brand.

For example, take a simple new employee press release. The typical press release is written with the expectation that it will be edited from the bottom up. In other words, PR people and editors know that the most important information will be at the top of the press release and they will typically trim from the bottom when there isn't room for the whole release. Knowing this, the press release should contain branding foundation information, like positioning or personality, near the top. Thus, a first sentence might read, "The Jones Company, the city's largest real-estate firm, announced today the hiring of Kimberly Smith to the position of realtor specializing in commercial properties." If a photo is included, Kimberly should be smiling, serious-looking, or somewhere in between, and be dressed appropriate to the company brand personality.

Public relations practitioners consider all of your "publics," or audiences, when planning a communications program. Your target customer, in addition to your vendors and distributors, are some publics to be considered during planning.

Public relations are not only an external function. Your employees are a very important part of your public audience and should be paid attention to with your public relations efforts. Creating "esprit de corps" or creating a sense of purpose and unity throughout your employee group is critically important to the success of your endeavor. Employee and human resources (HR) communications deserve the same attention to detail and brand consideration as communications to your external publics. Remember, your employees are key touch points in your company's brand experience. Treat them like the gold that they are.

What follows is a list of public relations initiatives in which a typical organization can engage. Most initiatives have a brief description of how they can enhance your brand and specific "Here's How" recommendations on how to do it yourself if resources don't allow you to hire professionals.

Public Speaking

Public speaking is an important PR delivery mechanism for brands because if you can impress your brand on an audience, each person there can help spread the word for you. There is an insatiable market of people wanting to know about most every business. Program planners for groups like the Rotary, Kiwanis, Chamber of Commerce, and professional groups of all types are constantly looking for entertaining and educational speakers. Taking advantage of this need is an excellent way to promote your brand. Talk to program planners. See what topics they are interested in. Write a speech that reinforces your company brand foundation and educates your audience at the same time. If you are uncomfortable talking in public, join Toastmasters and learn to be a confident speaker. Don't be afraid to speak from notes if they help. If you are uncomfortable speaking in public, write out your entire speech, practice it by first reading it aloud several times alone. Add personal stories that you don't need notes to remember. Once at the podium, the occasional glance at your notes won't be noticed.

Here's How

Start planning your speech by making sure that you know what your key message is and what you expect your audience to take away from your presentation. Don't forget to keep a focus on your brand foundation. In other words, be prepared to show your personality, confirm your position, and be true to your brand promise.

Back in college when we all took the required speech class the teachers all seemed to have one universal rule that is worth remembering. It is an old adage, which we have all heard, yet it is as true today as it was back then. It consists of three simple rules in one simple sentence about how to construct a speech: "Tell them what you're going to tell them, tell them, and tell them what you told them."

Another simple and effective device is to weave in stories. Everyone likes a story and they are easy to remember and learn from. Jokes can also work the same way but are dangerous and very hard to pull off, so be careful. Remember, even if your talk isn't about what your company does, people will be judging your company by what you say and how you say it. In this case, you are your brand.

Media Relations

Newsflash! The media are run by humans. Writers and editors are living, breathing, opinionated, flawed, and decent people with their own beliefs and values. The occasional press release isn't enough. Get to know the writers and editors that write about your industry. Ask them what they need from you to help them do their job. Be straight with them. Treat them with respect. Let them put a face to a name by sitting down with them.

Several years ago, our firm was hired by a new winery to help them develop a public relations program. In the wine industry, wine writers are critical to success. They can make or break a product. We knew it was important for our client that wine writers respected the brand. With the help of the winery, we located about 150 of the most influential wine writers in the country. We sent each of them a letter from the winery telling them that we were a new winery that was setting out to do things differently. We told them that we respected them as writers, knew that they were inundated with material and wine from thousands of wineries, and asked them to let us know how they wanted to receive information and what kind of information they wanted. We invited them to e-mail, call, or mail us their answers. To our surprise and delight, over half of them replied. We had instantly made both a connection with a group that is notoriously hard to connect with and built the beginning of a relationship with them.

Here's How

For proactive media relations, compile a media list of relevant publications and broadcast media. Specify the audience the medium serves and why that particular medium should be interested in you. Make sure that you have the names of the relevant editors or writers that might report on your business. When the list is complete, call and confirm the name of the relevant writer or editor. Also ask them how they like to receive information and what they think is relevant to their audience. Add all of this information to your database and then make sure you stay in touch so the reporters think of you when they have a story that relates to your field.

Crisis Communication

One brand-affecting public relations issue that companies hope to avoid is crisis communications. Yet, it is in times of crisis that a brand is under close scrutiny and in the public eye. Fortunately, there is a way that you can actually increase your brand value by reacting to crisis in a planned manner.

In 1986, Johnson & Johnson set the standard for crisis communications when they voluntarily recalled Tylenol from store shelves across the nation after someone added cyanide to some Tylenol packages. The company's quick action, and their subsequent invention of the tamper-proof package, was widely praised. Although the recall and reintroduction cost Johnson & Johnson a lot out of pocket, the public relations value was unparalleled.

Here's How

Crisis communications is a reactive media relation's initiative that should be thought of as much as possible in advance. Think out what possible issues might arise in your line of work and have a plan for how you'll deal with them. That way, you are not reacting but rather acting in a brand-positive way. You'll be creating your plan at a time when your thinking is clearer than it might be in an emergency situation.

The best advice for crisis communications is to take these four steps: tell the whole story, get the story out fast, tell it truthfully, and state what you intend to do about it. This approach will reinforce your brand promise and personality.

Key Messages

PR practitioners use research to determine attitudes and opinions of the public or the specific segment of the public that they want to influence. They use this research to help shape messages. You'll hear PR professionals refer to these as "key messages." These are the key points that they want to reinforce throughout their communication efforts.

Public relations initiatives are often used tactically to achieve a specific short-term goal. In other words, PR is usually used most effectively to announce something of immediate interest. In that context it is easy to forget to reinforce your key messages and brand foundation.

It is a good idea to develop your list of tactical key messages in the larger strategic context of your brand. In the context of branding, these messages should always reinforce as much of the brand personality, promise, and position as possible without cluttering the main tactical objectives.

Here's How

Your key messages are typically of two types: long-term strategic brand messages and short-term tactical public relations messages. For example, it may be that one of your brand messages, intended to reinforce your brand, is something like, "We are a fun, exciting company." Your short term PR message may be, "There's limited time to sign up for our summer special rates." These can be combined by using a fun, exciting name for the rates or using fun copy to explain the program. For example, the name could be "Rip Roarin' Rates," and the copy could use terms like "Sizzlin' Summer," and so on.

Events

Most companies, at sometime in their lives, stage some kind of event for their clients or employees. Events are another important distribution point for the brand and a golden opportunity to represent the brand on a personal level. Events can include trainings, sales meetings, new product introductions, award ceremonies, holiday parties, or grand openings. Even sales are events. Most events are themed. Events often represent missed branding opportunities because event planners are thinking of their event theme rather than reinforcing the company brand. Look for ways to make the theme reinforce your brand foundation from the very beginning.

Here's How

Start the event planning with a creative brief that outlines the audience, goals, objectives, key messages, brand expansion opportunities, and evaluation methodologies. Include a typical event project brief in your brand manual.

To use our company as an example, at Funk/Levis & Associates, Inc., we have, as part of our company brand, a value that we give back to our community. To demonstrate this commitment, at our annual holiday party for our clients and friends, an event that draws hundreds, we ask the attendees to bring either gifts or books for children or food for our local food bank, depending on which nonprofit organization is picked to be the recipient of our guests' generosity. It reinforces our brand, it gives people a reason to come to our parties, and it makes the attendees feel good about themselves. It's a win-win-win all around.

Community Involvement

One of the areas of brand development that doesn't get much press is community involvement. In this case, your community is whatever you define it to be. It could be your geographic community or your industry community. In either case, community involvement represents another of the many touch points of a brand. In many of the professional service businesses, such as attorneys, accountants, bankers, insurance, and real estate brokers, community involvement is a major key to brand development.

Charitable Giving

Most every company is asked to give back to its community at one time or another. Some companies, because of their size, charitable reputation, or visibility, are constantly asked for donations, auction gifts, or in-kind services. There are so many good causes, both social and political, that it is difficult to say "no" to requests for donations. Many companies have figured out that their charitable giving can reinforce their brand while simultaneously satisfying their very human need to be good members of society. What could be better?

Aligning your company with a specific nonprofit can work the same as Nike's successful alignment with Michael Jordan and other star athletes. The transference of the positive brand characteristics of the charity to your company can be as beneficial to you as your help is to the charitable organization. If your company makes or sells products for children, for instance, then participating with a nonprofit that aids children makes perfect brand sense. In the same vein, if you market primarily to women, being associated with nonprofits that are respected and admired by women only reinforces your brand.

When planning your giving, look for opportunities to link up with nonprofits that fit your brand. This is not to discourage you from contributing to other worthy but less visible causes. Give to them also. Give to them expecting no benefit apart from feeling good and helping those in need. That's probably the best benefit of all.

Political giving can be valuable or it can be bad for your brand. Because people are so passionate about politics, you risk alienating as many people as you make happy. However, if you market primarily to gun owners, for instance, then donating to the National Rifle Association might make sense. The same holds true with environmental organizations. Companies like Patagonia are well known and respected by their outdoor-oriented customers for their generosity to environmental causes.

Here's How

Write down a list of organizations that reflect your brand. Determine how you can be involved with them, be it charitable giving, volunteer release time for your staff, or in-kind donations of time or materials. Also, create a policy about how your name can be used in conjunction with the organizations and develop a contract that spells out your requirements.

Sponsorships

Sponsorships are similar to charitable giving. Sponsorships help leverage marketing dollars and work in the same way as charitable giving does in transferring brand capital. Sponsorships are also a good avenue for approaching new prospects or reinforcing your connection with existing customers is a different way.

Here's How

Create the same kind of list as is described for charitable giving. Remember that you are piggybacking on their brand and your objective is to transfer the positive benefits of their brand to your brand, not the other way around. Look for the intersections between your brand and theirs and brainstorm ways to reinforce this intersection. Perhaps you can do more than just add your name; for instance, you can add to the brand experience of the attendees by providing souvenirs.

When negotiating a sponsorship, make sure that you bargain for a high degree of visibility and that you have control over how your brand is presented. Create a contract that spells out your requirements.

Industry Associations

Involvement in your industry, aside from the benefits of information sharing and advancement, is good branding. Being considered an industry team player helps blunt criticism from the competition and builds credibility for your organization. This is especially true in the professional service trades, where credibility is what you sell. In addition, if you belong to an industry that is generally well respected by the public, being connected to an industry organization reinforces your brand.

Community Boards

Most service businesses know that involvement on community, state, or national boards of directors, such as the Chamber of Commerce, the Rotary, and in-service and social nonprofits, is good business. It's good business because it allows your employees an opportunity to meet other professionals and business people that can help your company grow. It's also good branding. It publicly demonstrates aspects of the position, personality, and sometimes the promise of the company to an audience that you otherwise might not reach.

Here's How

Don't assume that you are the only one in your company who can donate time or expertise. Encourage your employees to participate. "Donating" employees and their time to a board of directors is much the same as giving cash. It is another form of in-kind donation. And it has the side benefit of letting more people know the kind of folks that work for you. If you donate employees, remind them that they are representing your brand and all the people that work with them.

Style

Voice

Voice is something that many companies don't consider, yet, it is an important component of personality. Your company voice is *who* is doing the talking for your company. Is it the management? The employees? The company voice should match the company personality. If your personality is formal, then your copy should reflect it. Consider if you want to talk in the first person ("We are the company . . .") or the third person (Smith & Jones is the company . . ."), then stick to that style.

Tone

Your tone is *how* you are talking for your company. Copy tone can be informative, persuasive, humorous, conversational, casual, formal or professional, scientific, or any number of different ways. Obviously, you want to use a tone that resonates with your customer base while still reflecting your brand.

Here's How

Determine the voice and tone of your company based on your personality description and your representative icon. Determine how conversational you are going to be in your communications and write down your standards for conversational tone and voice in your brand document.

Keywords

Keywords are different than key messages. They are words, often adjectives, that are brand related and are used in your copy for all of your marketing materials and public relations efforts. The objective of using keywords is to connect your company with those words through repetition.

Here's an example of choosing the right keywords. Several years ago, members of our firm were involved in developing the strategy and marketing materials for a statewide campaign to add hundreds of miles of rivers to the state scenic waterways system. Our objective was to stop mining, dam building, and timber clear-cutting along the banks of key waterways. Our mantra as we were building coalitions was "no dams and mining," because that was what we wanted to prevent. Environmental groups, rafters, kayakers, and fishing groups signed on to the campaign because they shared our concern about the negative effects of dams and mines on rivers. We all agreed they were environmentally harmful.

Rather than immediately using the concept of "no dams or mining" in our marketing communications, we conducted research among likely voters to gauge their reaction to the "no dams or mining" message and other root concepts that related to the issue. It turned out that, to non-river enthusiasts, dams and mining were not even a concern. The words that resonated with the majority of likely voters were "fish" and "future generations." That turned out to be valuable information. From that point forward, we hammered on the concept that we were saving "fish and future generations" everywhere we spoke and every time we created marketing communications. The result? We won the election handily, even though we were battling a coalition of resource extraction industries that had 10 times the campaign budget.

Here's How

Once you have chosen your keywords, create a list of those words to use in copy and in conversation. Try to limit the list to as few words as possible. As many of these keywords as practical should be used in all of your communications. Check the relevance of these words with some of your customers through formal research or informal conversations.

Tradeshows

To a high degree in the manufacturing and distribution industries, and to a lesser degree in most other trades, a tradeshow is an important brand touch point. By now you understand that your tradeshow exhibit must reinforce your brand foundation. But you'd be surprised at how many companies don't get the connection. Tradeshows are a rare opportunity to reach out to interested potential customers as well as to reinforce your relationship with existing customers. You and your brand should be at your very best.

Think of a tradeshow as a medium, just like you'd think of radio as a medium. Every medium has its own particularities. In a typical trade show, there may be thousands of booths exhibiting a wide variety of items. A major objective is to be noticed. Another is to be inviting.

Here's How 1

First and foremost, make sure that your tradeshow booth makes it obvious who you are and what you are trying to tell them. In a crowded and confusing environment, it is imperative that you are as clear and obvious as possible about your messaging. Make sure that your booth feels like your brand and that your graphics feel like your ads, your collateral, and your identity. Make sure that you take the opportunity that a tradeshow affords to visit one-on-one with people, make them feel welcome in your temporary home and promote your brand. If you have giveaways, make them reinforce your brand. Use bright lighting. Make sure access into your space is easy. Don't put a big counter between your booth and the aisle. Also, make sure that your booth clearly shows, at a glance, what it is that your company does. Tradeshows tend to be highly visually and aurally cacophonous. Make sure you stand out.

Here's How 2

The primary objective of a tradeshow should be to build relationships rather than sell products. Tradeshows are an opportunity to visit and cement relationships with existing customers and to create new relationships with potential customers and the trade media. They are also an opportunity to see what the competition is up to. So, plan accordingly. Start with a creative brief. Design your tradeshow booth to brand standards. Train your floor staff and remind them that their job is to build relationships, not hand out literature.

Web Site

Everyday more and more people are utilizing the Internet to investigate and compare companies. Your Web site may be one of the first places or touch points for you customers to encounter and experience your brand. They will make judgments about your company based on ease of use, load time, graphics, content, and the "cool factor" (is it cool, engaging, and interesting). Your Web site is a critically important touch point regardless of whether you are a "Web-based" business or not. If you are a broad-based marketer, or marketing to a nontechnical market, make sure that your Web site is user-friendly, works with the slowest connections, loads fast, doesn't rely on users having specialized software and is not graphics-heavy. If you must picture your products, make sure that the files are not too large. We won't attempt in this book to go into all the aspects of Web design, but we will caution you to make sure that your Web designers understand your brand and take advantage of the medium's attributes to reinforce your brand foundation.

Here's How

Before you build your Web site, check all your competitors' Web sites. Remember, you want to differentiate yourself, so it helps to know what you are differentiating from. Again, with every marketing communication vehicle, start with a creative brief. Then create a site map that shows how your Web site will work. A site map (see Figure 8.13) is really like blueprints for a house. A site map tells you how many pages you will have and how the information on your Web site will be organized. It isn't difficult to do, but you may want to have a professional Web designer help you. Once you have your site map, design a look that reflects your brand and is consistent with all other materials. This is especially important on the Web. Remember that the Internet is not just a place to provide information like an online brochure. It is a place to share information about your brand and the brand experience and a place to provide value to visitors. Value is important as the Internet grows up (i.e., Web 1.0 to Web 2.0 and even 3.0). If these terms are not familiar to you, Wikipedia has an extensive description of what they mean. Any Web company should also have ideas and information that will make your new Web site more engaging and useful.

When you have the site map and the overall look and feel of the design then your next step is to build the pages. Sounds simple, right? Not quite. Web development is an iterative process and you'll find yourself making compromises as you move forward. It really is best to have a professional guide you through this process.

Make sure you hire a Web designer that is comfortable with Web 2.0 strategies and social media connections (see next section for more on this idea).

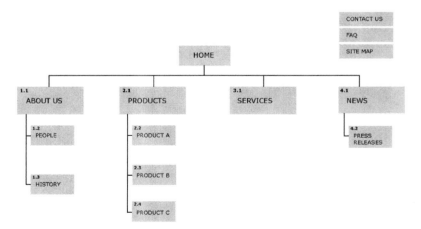

Figure 8.13.

Social Media

How do you define a medium that is still defining itself? In as few words as possible. By our definition, social media are a way of sharing information and connecting with communities via tools such as Facebook, Twitter, YouTube, Digg, Flickr, and many more. Social media are a revolution in the way information is shared via the Internet. It may seem as if this is just a personal thing and not a business concept, but we disagree. Social media will transform the way that companies interact with their customers and will allow customers to give feedback on *everything* in real time. In the business world, you will hear concepts thrown around like consumer generated media (CGM) and user generated content (UGC). What you really need to know is that this will this affect your business in a very fundamental way. If the Catholic Church is considering using Twitter as a way to get the word out and gather information, shouldn't you be thinking about it too? So, what can you do to get involved and ride the wave of social media?

Here's How

Spend time researching social media. There are millions of sites on the Web, but here are some of our favorites: Wikipedia (one of the original user generated content mediums as the whole site is generated by everyone) and Twitter. Twitter is a fantastic tool for researching almost any subject. It has a fantastic search tool and with as many users as they have, there is information right at your fingertips. Searching the general Web is also a good way to gather information. The best way to understand the medium, though, is to use it. If you don't know how to get on Facebook, LinkedIn, Twitter, or any of the others, ask one of your employees. Chances are, they are connected in ways you never dreamed of.

CHAPTER 9

Product/Service Design

One of the roles of marketing communications is to set expectations. As we mentioned earlier, when these expectations are not met, there is a brand disconnect. A brand disconnect makes people wary and confused, or worse yet, disappointed. If you disappoint your market, then you have actually used your marketing dollars to harm your enterprise. Since that is obviously not good business, it pays to make sure that your products and services are reflective of your brand.

Let's look at a typical company to see how you might easily create a brand disconnect. Manufacturers often have layers of brands. They can have the corporate brand, one or more category brands, and product brands (see chapter 11 for a discussion on corporate, category, and product brands).

For example, consider LWO Corporation, a respected manufacturer of wood products, based in Portland, Oregon. LWO has two divisions: Woodway, a manufacturer of wood products for professional contractors and carpenters; and Arboria, a manufacturer of wooden outdoor, yard, and garden structures aimed at a consumer market. In this case, LWO, as a corporate brand, is almost invisible to the public, yet known to the distribution channels. Having a strong corporate brand such as LWO's makes it easier to launch category and product brands into retail and distributors.

Woodway, the company's original brand, is a category brand. It has a brand foundation based on the position of leadership in quality construction and a personality based on the down-to-earth, no-nonsense styles of LWO's founders, all of whom were professional carpenters. Their promise is that they will always provide reliable, flawless, field-tested products. They emphasize this with their trademarked "No Call Backs" tagline and accompanying symbol that appears on their ads and brochures.

Call backs, which are the calls that contractors get from customers to fix something wrong in the project, are the bane of professional contractors.

As stated earlier, Woodway is a "category brand" of LWO Corporation and positions itself as the leading producer of solid wood, overengineered, high-end, intelligent, professional-grade products for discerning contractors. It promises a guarantee of complete satisfaction. Under the Woodway category umbrella are several product brands such as Architectural DeckRail, EZ Rail, Classic Post Caps, and others. DeckRail itself has individual product styles or sub-brands including Council Crest, Irvington, and Laurelhurst.

DeckRail is positioned as the fast and simple way to create deck railings, a labor- and cost-saving solution, and as expected of the Woodway brand, a high-quality and intelligent product. The product personality is professional, beautiful, and natural. The styles of DeckRail are not positioned differently than the product brand.

There is a continuum from the corporate brand to the product brands that makes sense. LWO is a manufacturer of quality wood products. What wouldn't make sense would be if LWO started a category brand or product brand that specialized in plastic molded materials. There would be a disconnect from the corporate brand that would undermine all the products of the corporation and the categories. Should LWO ever want to branch out to plastics, they would need to change their positioning from wood products to products for the home and garden.

Let's look at some examples of applications of brand thinking to products and services and examine a process for developing brand-related products and services.

Product Design

How do you develop new products or services that are related to your corporate brand? One way is to look at your corporate or category brands and see how to transfer their brand foundations to a product level. What are the personalities, positions, and promises of your corporate or category brands? Do those traits suggest product or service names or ideas? Do they suggest ways to redesign your product or packaging?

Sustainability is another issue in product design. Can the product be recycled? Is it made and packaged in as earth-friendly a manner as possible? Cradle-to-cradle manufacturing (where manufacturers are required to take products back at the end of their life cycle) is already regulated in parts of Europe and will soon likely be standard worldwide as triple-bottom-line sustainability gains traction. From a business standpoint, triple bottom line sustainability means that the business is profitable, that the business takes care of its people, and that the business is good to the environment. If your brand includes values that relate to responsibility, environmental or social values, or people-centered values, sustainability will be a brand criterion.

Here's How

Most products are designed to perform a function. Often these products are designed by engineers or product designers who are focused on function more than they are on brand considerations. Increasingly, however, product designers are aware of the importance of connecting the style of the product or the packaging to the company or category brand. If you are a manufacturer, you should examine your products with a product team and consider whether you could design them differently to more closely align with your brand foundation. Include in this audit issues like the color, shape, and tactile feel of the product. Apply your brand personality attributes to the product. Think about the use of shape and typography. Think about the product name. Think about your brand.

Service Design

Having services "packaged" gives customers a sense that you have done this before, that you have an established methodology, and that the outcomes might be predictable. Bundling, or packaging existing services, and then naming the package can be an effective differentiator. An example might be that as a home insurance agent you provide not only insurance but also home safety audits as a value-added differentiator. Name your package the "Home Aware" program and suddenly you have a branded program that is exclusive to your business.

Here's How

If you have standard operating procedures for your service offerings, which is a good idea if you are in a service business, then you have the makings of "branded" services. If your service is essentially the same as that offered by competitors, look for simple ways to differentiate the service, such as adding an extra step or some follow-up procedure. Then, pick a name, and maybe a logo, and suddenly you have an exclusive product that sets you apart and may have higher perceived value to the customer.

Packaging

Packaging is often the easiest way to spruce up a product. It is usually easier to redesign packaging to reflect your brand than it is to change the product inside. The Japanese, with their extravagantly wrapped gifts, and perfumers, with their costly packaging, have known this for decades. In fact, quite often, the cost of the packaging for perfume exceeds the cost of the contents. Packaging is critically important to branding.

Over the next decade, we expect packaging to gain prominence in the public consciousness. Recyclability, minimal packaging, recycled materials, and ease of use issues are all packaging considerations being examined by designers and manufacturers worldwide. Over the years, packaging has evolved from being merely protective, to being a critical step in the sales and branding processes, and then to being designed for safety and antitheft considerations. The next evolution will add sustainability to this list. The important thing to remember is that all these packaging issues have brand implications.

Here's How

If you are a manufacturer, look at the product packaging and shipping materials you use from a perspective that all the functions of packaging are possible opportunities to reinforce your brand. Look at each function individually:

- Protection
- Point of sale
- Safety
- Ease of opening
- Antitheft
- Sustainability
- Instructional
- Brand identification

Then, ask your brand team how you can make each function consistent with your brand.

Retail Products

Bundling

If you are a retailer, you have the power to bundle or package products to, in a sense, create a differentiated product that reflects your brand. For instance, bundle a plastic spoon with a container of yogurt and you create an instant snack. Or in a clothing store, by selling a shirt and sweater as a packaged look, you create an ensemble and potentially increase your sales. No matter what kind of retailer you are, look around your store and see if there are any potential bundles you can create. Bundling products in retail is also a concept that can be applied to restaurants, such as pairing certain wines with certain entrees, or creating the equivalent of McDonald's Happy Meals.

Unique Products

Having exclusive rights to market products in a specific geographic area is beneficial to your brand if those products are reflective of your brand. Look for products that are unusual and unique and feel like your brand.

Sensory Equity

In an effort to differentiate themselves, savvy retailers are looking beyond visual differentiation and exploring how they can create a sense of brand through all senses. New concepts like sound equity, tactile equity, and scent equity are now considered by manufacturers and service businesses alike as a branding step.

One good example of sound equity is Harley-Davidson's attempt to trademark their motorcycles' unique, deep-throated sound. In 1994 Harley-Davidson submitted a trademark application with the following description: "The mark consists of the exhaust sound of applicant's motorcycles, produced by V-twin, common crankpin motorcycle engines when the goods are in use." This was not the first time a company had registered a sound. For instance, company jingles are usually registered. The MGM Lion's roar is registered. But Harley eventually withdrew their trademark registration attempt, winning in the court of public opinion but getting nowhere on the legal trademark.

Perhaps your company sound could be the type of music you play in your establishment or the sound of an indoor fountain trickling through rocks.

Scents are a particularly powerful and evocative sensory experience. According to Harvest Consulting Group, LLC, a company specializing in building brands through sensory experiences, certain scents have immediate associations. They use examples showing the association of the scent of lemons or pine with "clean," baby powder and lavender with "nurturing," leather with "tradition," and salty air with "adventure." Harvest Consulting points to various studies dating from 1935 that show that memory for odor stays with us over time, is easily accessed, and has a high degree of clear emotional association.

Retailers and manufacturers are increasingly paying attention to scent equity and touch equity as brand touch points. Retail stores are releasing scents into their establishments; furniture stores are baking cookies for customers inside their stores to evoke fond memories of home; used car dealers are spraying the new car smell into used cars; and manufacturers of all types mix smooth woods and sumptuous leather in products. Is there an opportunity to evoke a desired emotional response in your organization through sensory equity?

Comarketing

Building brands can take time and be expensive. To reduce costs, manufacturers, retailers, and service providers comarket or copromote their products and services. Often you will see multibrand manufacturers cross-promote their different brands, for example, by offering a discount on their snack brand when you buy their drink brand. Comarketing is also often done by two different companies that market to the same audience, such as a snowboard manufacturer and a snowboarding apparel manufacturer. It happens at the retail level, both internally, such as when a retail store offers a discount on shirts when you buy pants, and externally, when two retailers copromote, such as when they are in the same mall.

The upside of copromotion is that expenses are reduced and you are possibly exposed to a whole new customer base. The downside is that you have no control over your comarketer's brand and how it can affect your brand. Therefore, there could be risks involved that you should take care to avoid.

Here's How

Before entering a comarketing agreement, examine all the potential risks to your brand. Develop a written agreement with your comarketing partner that addresses how those risks will be addressed and outlines the responsibilities and liabilities of each partner.

Brand Extensions

In our decades of business, we've seen many companies paint themselves into a corner with brand extensions. We've seen companies cannibalize their own existing brands, confuse their customers, and run out of extension name possibilities. Usually these problems exist because of naming issues that weren't anticipated when the original name or first extension name were developed. As you are now aware after reading this book, naming is a critical brand issue. Fortunately, these problems can be mostly avoided with preplanning.

Here's How

The trick is to plan ahead. When you develop a new product or service, even if you think it will be the only one of that type you ever do or that you'll never improve on the product or service, plan your possible naming extensions in advance. If you do improve on the product or service, or change it slightly for a different market, you'll be ready and not find yourself trapped in a difficult brand extension issue. This preplanning also turns out to be helpful should you purchase a competing product or company and want to absorb their products or services into your line.

CHAPTER 10

Operations

Operations covers all aspects of operating your business, including such functions as payroll, human resources management, purchasing, and a myriad of other operative processes and procedures. Operations is usually not thought of as a brand building arena. Quite often the marketing department and the operations department do not see the relationship between their two functions. Yet it is in the area of operations that many companies stand out as brand builders. For example, look at companies like Oil Can Henry's. Every one of their buildings is of the same distinct design, with tall glass fronts and their distinctive red and white color scheme. Or look at Les Schwab Tires, the famous western state tire company whose employees run, not walk, to your car when you pull up for service. Or look at the Ritz Hotel chain, where every single employee, regardless of their job function, is a welcoming host or hostess.

Let's look at four aspects of operations to see how they can affect branding.

Facilities

In certain types of businesses, facilities play a critical role in branding. Take, for example, a restaurant. The exterior of your building, which is the first impression a potential customer has of your business, can say a lot. You can express—through design, color, and a host of cues—what a customer can expect. The typical customer can quickly, and seemingly intuitively, make a guess as to your pricing, the level of service one might expect, the way they might be expected to dress, whether or not you serve alcohol, or if it is an appropriate place for young children. The customer makes all these judgments at a glance, based on past experience.

Once inside the restaurant, the patron is engaging more deeply with the brand experience. The interior either reinforces first impressions or contradicts them. Ideally, it's the former. If the customer's experience of your brand is consistent and familiar feeling, the customer will be comfortable. If that's the case, your facilities have started a relationship with a prospective customer even before you've engaged them person-to-person.

Facilities say a lot about a company and go a long way to building expectations about cost of products and services. You wouldn't expect to pay $300 per hour to a lawyer working out of a shabby, run-down office, but you would expect to pay that for one working out of a well-kept, plush office filled with original art. Because the latter attorney's office "feels" successful, you'd have more confidence it that firm's ability to solve your problem. So, even before sitting down with an attorney, you have begun a relationship of confidence based on brand cues broadcast by facilities.

Here's How

If you are building new, or remodeling an existing space, go back to your brand. Remember that colors, shapes, lighting, and every other aspect of your facilities are symbolic representations of who you are. Make sure that your architect is aware of what you want to express to visitors and employees. Share your brand foundation with the architect and interior designers and ask them how they think that you can reflect your brand through their designs.

Hiring

Having a strong company brand affects hiring in at least two ways. First, good employees are attracted to good places. If you have done a good job of determining and then marketing your brand, good people will be attracted to you for the same reason that good customers will be attracted to you: confidence. They'll feel confident that they will have secure employment and that working for you will reflect well on them.

Second, a brand helps you determine hiring criteria. If you know your brand well, you'll hire the kind of people that will enhance it.

Here's How

When interviewing prospective employees you should try to determine how likely they are to live your brand and reflect your brand to customers. Develop lists of questions that get to your core brand values and words. Do their values align with your company's values? For instance, if you value honesty, ask a prospective employee for examples of how honesty has and has not paid off for them in the past. Have them give real-life examples as opposed to mere assurances of their honesty. You can also ask them to list their core values, or who their personal heroes are. A good question is to ask them to describe what one thing they are proud of in their work history. This will help you to determine if they will align well with your brand. Ask them what they think your brand stands for. You'll learn a lot about them and about your company by doing so. Ask them for ideas about how you might make your company brand stronger. Ask them for the name of another company that seems to share many of your company values. Let them know, during the interview process, that brand is important to you.

Another good way to interview potential employees is to create an interview team (this could be part of your brand team responsibilities). Having a team of people interview potential employees gives you different perspectives on how someone fits your brand. It also increases employee engagement in the process and may add to the success of bringing new team members into your company.

Training

It's one thing to build a brand. It's a whole other thing to maintain it. Training is an important part of brand management. As we've said, employees are a primary brand touch point. To make sure that they are fully participating in advancing the brand they need to understand expectations and be trained to meet or exceed them. Training should always emphasize brand to a great degree, even if a specific training exercise doesn't appear at first glance to involve brand. Thus, training on new software can include information on how the software can enhance the brand experience.

If your company is large enough to have a human resources (HR) manager, the HR person should be an unabashed brand champion. In many cases, the people hired and trained by the HR department are the absolute keys to a successful brand experience for the customer. If you've never stayed at a Four Seasons Hotel, it's worth the investment to spend a night in one just to see what a difference a superbly trained staff can make on the visitor experience.

Here's How

Regardless of the business you are in, new employees almost always need some training or orientation. Even experienced people in your field, when moving over to your company, need to understand and be trained in your brand. Once you've hired employees, give them a copy of your brand manual and expect them to study it. After a week, ask them their thoughts on the brand and perhaps even test them on it. Let them know that living the brand and contributing to its success is a surefire way to rise in the company hierarchy. Of course, if you make such a promise, be prepared to honor it. Have ways to determine their brand performance. Reward them publicly for small steps.

Remember that training, like any other business function, needs to be maintained. Hold brand-training sessions annually. Ask for employee input into how to enhance the customer brand experience.

Procedures

In the brand map we have extended the "procedures" section into the next level because there are so many operating procedures in any company that it is difficult to address them en masse without isolating a few to use as examples. Our office is typical. We have procedures for entering jobs, doing the work, and billing clients. We have account management procedures. We have client service procedures, identity design procedures, and procedures for serving coffee to our guests. Each of these procedures is a brand-building opportunity. Let's look at a few typical company procedures in the context of brand.

New Client Intake

Remember when we were discussing "facilities" and we used the restaurant example? We demonstrated how facilities could create expectations and establish brand. The same is true with new client or customer intake. For example, in our hometown there is a company that changes the oil in your car. The sign on their building promises a 10-minute oil change. When a customer pulls in, a man jumps up and guides the driver into a bay. With a smile he comes to your window and asks what service you need. You request an oil change. He quickly fills out a form with the oil type and the car model. He then asks you how many miles you drive per month. Then he asks you to fill out the customer information section, which consists of two or three lines for name and address, and so on. While you are doing that he checks a manual for the oil quantity your car needs, then pops off your air filter to check it. If it's clean he tells you. If not, he shows you and recommends a replacement.

While all this efficiency is happening, a man in a bay beneath your car is draining your oil. The man up top is checking and adding to your radiator and windshield fluids. The man below signals that the car is drained and the oil plug replaced. The man up top immediately pumps the required amount of oil into the vehicle. Then he checks the transmission fluid. In the meantime, the customer is dazzled by the efficiency, feeling like a racecar driver at a pit stop. Finally, in far less than 10 minutes, you pay for the service and receive a free car wash and vacuuming of your vehicle, an unexpected bonus the first time you visit, then an expected reward for your loyalty on subsequent visits. The subsequent visits are almost assured, because in a few months, based on how much you said you drive, you receive a postcard reminding you that it's near time for another oil change.

The point is that during the client intake process, the driver learns that key components of the brand of this company are efficiency, friendliness, service, and customer care. The company reinforced their brand components through the intake process.

Here's How

Examine your intake process from the point of view of the customer. See if you can find out your competitor's intake process, if possible. Take what you learn and establish a protocol for how you will intake new customers. Ask current customers about your current intake process and if there are suggestions for improving it from their perspective. The key is to make it as easy as possible for the customer to do business with you and find ways to bring your promise to life.

Sales Process

Since the sales process is often where customers first come into personal contact with your business, it is a key component of brand building. The sales process should be designed to reinforce your brand. How you integrate the brand into the sales process depends on the nature of your company. To demonstrate, we'll describe brand-infused sales processes in a service company and a retail business.

We'll start with a service business. Let's say you are responsible for new business development for an accounting firm. You have determined that the personality of your firm is one of quiet professionalism. The position that you can claim is that you are the leader in efficiency and accuracy. Your brand promise is that you are always honest and that you will explain things in layman's terms.

How the branded sales process works depends on how the connection was made. Clients can either connect with you through a referral, through the Yellow Pages or the Web, via your sign, an ad, or a chance meeting. Or perhaps you made the first move by making a cold call. Since the latter is usually the hardest, we'll use that as our example. This is not an attempt to teach you sales techniques, but rather, to show you how brand can be infused into the sales process.

You start the process with a letter to the prospect. The letter, on your professional letterhead that reflects your brand, is addressed directly to a person ("Dear Mr. Smith" rather than "To whom it may concern"). It is polite, expresses why you are contacting him, and says who you are and why you're different than other accounting firms. You might mention how you got his name and list some of your current clients. At the end of the letter, you promise to call on a certain date to talk.

On the promised date, you call the prospect's office and ask for him or her by name, informing the receptionist that he is expecting your call or that you are calling in follow-up to a mail correspondence. When the prospect answers, you thank them for accepting your call, promise to take only a few minutes, and explain why you can make their life better. Your objective in the phone call is to get a personal meeting, not a sale. That comes when the relationship is established. At this point, you are establishing your brand by building a relationship.

If you are successful getting a meeting, the main objective is to show how your brand differentiates you and is superior to his current accounting services. You should have found out in the phone call what he wants from an accounting firm. When the meeting is completed, you thank the prospect, promising to stay in contact. When you return to your office, you immediately send a handwritten note of thanks.

Whether or not you make the sale, you have demonstrated your brand. Even if the prospect has a long-standing, satisfactory relationship with their current accountant, you might be potentially referred by the prospect when a colleague asks for a recommendation, or you may be first in line when the incumbent accountant makes a mistake. It is important to remember, even if you don't make the sale immediately, to maintain the relationship over time. The effort almost always pays off.

For our second example, a retail environment, we'll use a gift store. This example could apply, with some variation, to almost any retail establishment. A key difference between retail sales and service sales is that in the latter the sales process is typically longer and the buyer is buying the service provider's credibility. In retail, the process is typically a shorter cycle, assuming the cost of the item is not too expensive, and the object being bought usually already exists. Automobiles and other high-cost household items are obvious exceptions to the short sales cycle. Therefore, the sales process for them is often closer to the sales process for services.

As in our other examples, the customer is already making assumptions about your brand before they walk in the door. For our purposes here, we'll assume that you have already ensured that those first impressions reinforce your brand. If it is a new customer, when they enter your store, they won't know where to go immediately. If your brand is about efficiency, you'll ask them if there is something specific you can direct them to. If your brand is about creating a pleasant shopping experience, you could ask them if they want to browse or if there are suggestions you can make to help them. You might offer them coffee or a glass of water.

Because there are so many types of retail stores and variations in brand, we won't go into step-by-step detail of the retail sales process. However, we will remind you of two things. First, have a written, brand-inspired sales procedure that you and your employees follow. Second, the sales process extends through the checkout procedure and into any post-sale follow-up with the customer that makes sense. Remember, you are not merely making a sale; you are creating a relationship with a customer.

Complaint Handling

Sometimes the truest test of your brand is how you handle an unhappy customer. It is at this point, when a customer complains, that you can either cement a relationship or send an ambassador of negativity out into the market.

Customer Experience

Everything that we have talked about throughout this book is about the customer experience. In our business we call it the "brand experience." We repeat it here because it bears repeating often. The experience the customer has with you is critical to the success of your branding efforts and the maximization of your resources. Always look at your business through the eyes of your customers. If you are not able to give an unbiased customer point of view to your own business, and most of us aren't, hire outsiders to do it for you. Nothing is more critical to your success.

Here's How

We'll address two kinds of complaints, solicited complaints and unsolicited complaints. Remember that, quite often, dissatisfied customers don't complain to you. But they might well complain to others. So it is very smart of you to give them an opportunity to complain by asking them if they were satisfied with the service or the product. This is a solicited complaint. Let them know you care about them. If they have a complaint, don't defend. Instead, acknowledge their feelings and rectify the problem immediately.

You should have planned in advance how you will handle unsolicited complaints. Complaints can be about products, services, employees, delivery schedules, or a host of other issues. Write out your complaint handling procedures in every instance you can think of. Those procedures should especially reflect your brand promise. If you have written procedures, employees can handle complaints on the spot rather than sending the unhappy customer up the management chain, which typically only makes the customer angrier.

CHAPTER 11

Brand Structure

Brand Structure

When you only have one branded product or service, brand structure isn't an issue. But, if you have—or start to offer—more products and services, it is wise to have a predetermined structure in place to accommodate the new products and avoid confusing your brand. We have worked with many companies that didn't consider brand structure at an early stage of product development and had to create structure as the subsidiary and brand portfolio grew in a willy-nilly fashion. Determining a brand structure is much easier before you have developed a product line or before you develop category brands. Doing it the other way around is like building a house, then calling an architect to come fix the design.

There are three primary brand structural models. There are various names for them, but we'll stick with monolithic, multibrand, and endorsed. There are pros and cons to each model. There are also combinations of these three primary models. In fact, quite often, because of acquisitions and product line extensions, companies have a combination of two brand structures.

Monolithic Brand

The monolithic model is similar to what Harley-Davidson has. Motorcycles, cigarettes, clothing, and other products all sport the same brand and logo. Harley-Davidson targets "outlaws" or outlaw wannabes, and their brand is easily transferable to many products that appeal to the same demographic. Advantages of the monolithic structure can include faster brand building, lower brand building costs, and a stronger barrier to entry. One clear disadvantage is that a bad experience or a decline in popularity of one aspect of the brand can affect the entire organization. In other words, if Harley-Davidson Cigarettes are involved in an unpopular lawsuit, the bad publicity can affect the whole brand.

Multibrand

The multibrand model is similar to Procter & Gamble's. P&G has a plethora of product brands, some competing with each other. Unlike Harley-Davidson, each individual product brand is more dominant than the corporate brand. Therefore products like Pampers and Luvs diapers and Crest and Gleem toothpastes compete on the store shelves with most consumers being unaware that they are owned by the same company. The advantages of this model are that it allows companies to very specifically target market segments, and it competes against itself rather than some other company doing so. With multibranding, the decline of one bad product doesn't affect the other products. The major disadvantage is that it is the most expensive to market.

Endorsed Brand

The endorsed brand model is similar to Nabisco, a division of Kraft Foods. Nabisco "endorses" each of their individual product brands with their familiar red symbol tucked neatly in the upper left corner of the packaging. This becomes the equivalent of a seal of approval that signifies that these brands are good enough to belong in the respected Nabisco family. This model combines some of the advantages of the other two. It is easier to add and remove products without affecting other products but, on the negative side, a bad experience with one product can affect the perception of the endorser and lower the endorser's credibility.

Consider which structure will best serve you over time. There are pros and cons to each model that vary by situation. You can determine what those are for your company by consulting with your own staff or with professional marketers.

Internal Brand Structures

Within multibrand and endorsed brand, there are often internal brand structures. For example, many companies have a three-tiered brand structure that encompasses the corporate brand, the category brand, and the product brand. Each tier targets a different market segment. You can see how this works by examining Johnson & Johnson, the giant health care product company. Johnson & Johnson has a corporate brand. This corporate brand is designed to both "endorse" individual categories and products and to appeal to the financial community. The category brand of Band-Aid, a Johnson & Johnson brand, is designed to transfer equity to all Johnson & Johnson product brands within the category and to appeal to retail purchasing agents who are looking to fill out their slots on the retail shelf with reputable products from a single distribution network. Individual product brands, like Band-Aid bandages, can be imbued with personalities that reflect the market, like cartoon Band-Aids for children and clear, businesslike Band-Aids for adult consumers.

Another interesting, but complex, example of three-tiered brand structure is Kraft Foods, the international food giant that owns household-name brands such as Starbucks, Kool-Aid, Oscar Mayer, and Taco Bell, among many others. Kraft Foods utilizes a multibrand strategy as their external brand structure except within their own branded Kraft product line. Let's look at their structure.

Corporate Brand

Corporate brand is the overall brand of the company. The visual manifestation of Kraft Foods, its logo, is the word "Kraft" in blue on a white field encircled by a red, slightly angular oval. The corporate colors are red, white, and blue. While this logo appears on the Kraft brands packaging, it does not appear on the category brands, like Nabisco, another well-known Kraft brand. The corporate blue of Kraft Foods is infused throughout the Web sites of the company's brands so there is a loose, symbolic association with the corporate parent.

Category Brand

Category brands are those that are niche-specific, like snack items. Kraft is in five primary categories: beverages, convenient meals, cheese and dairy, grocery, and snacks. In the snack category is the Nabisco brand. The Nabisco logo is the familiar red triangle in the upper left corner of the products with the oval circle and strange antenna-like extension.

Product Brand

Product brands are the individual brands within a category. Within the Nabisco category brand is the product brand, Triscuit, which itself has product extensions like Triscuit Rosemary & Olive Oil, Original, Reduced Fat, Cheddar, and other flavors.

CHAPTER 12

The Other Side of the Relationship

Never forget that building a brand is building a relationship. In any relationship, it obviously pays to know the other person. Relationships aren't all about you. They are also about connecting with your customers. Your customers need to relate to you and you need to relate to them. So, any good program begins with an understanding of who your customers are, what their needs are, and how you can satisfy them. Marketers are constantly doing consumer research to probe the psyche of their markets, trying to find out how to motivate them, and trying to determine what words and images connect at an emotional level.

Knowing your market is a tall order. Helping you know it is a big industry. Every city has market researchers whose job it is to learn about what motivates people to buy, vote, or entertain themselves. Good market research examines not only what people think and feel but also what factors in the environment affect why they think and feel the way they do. These factors could include income, age, marketing communications, public relations efforts, lifestyle, values, and more.

Entire shelves of books have been written about market research. Formal research is one of the primary means marketers have of knowing their markets. Of course, one other important way that marketers learn about their markets is in their daily face-to-face or phone-to-phone involvement with their customers. That direct contact is considered informal marketing research.

The craft of formal marketing research is constantly changing. Standard research instruments such as focus groups, mall intercepts, phone, Internet, and mail research are infused with new techniques and the experience of psychologists and psychiatrists. This book only deals superficially about knowing your market. We leave that educational task to others more qualified. Our focus is on knowing yourself (your company) and understanding how to project yourself through channels.

It does pay to have a basic understanding of market demographics, geographics, psychographics, and technographics because you must ensure that you are relevant to the market. Obviously, you do not operate in a vacuum.

Demographics are the description of your customers from a statistical point of view. Demographics describe people in terms of data such as average age, income, cultural background, and education. Demographically you might describe your typical customer as a working 35-year-old married woman, college graduate, with two kids, two cars, a household income of $54,000, and home ownership. Or, you might say that the demographics of Southern California are changing as the Hispanic population increases.

Geographics is often considered part of demographics, but it is sometimes worthwhile to separate geographic data out of demographic data. Geographics is simply the study of where your customers are. The story earlier in the book about the coffee kiosk is a good example of understanding the local geographics of your customers. Another example of geographics is the idea that different sections of the country have different lifestyles and ways of thinking. *The Nine Nations of North America*, the seminal book by Joel Garrieu, discusses how different the people are in the various parts of the United States. The same concept can be applied to states and cities. For example, precinct voting patterns are a good indication of how people in various parts of your city think and act differently.

Psychographics describes how people seek experiences and buy products and services. The VALS system and corresponding surveys (originally known as "values and lifestyle") by SRI Consulting Business Intelligence (http://www.sric-bi.com/VALS) provide definitive psychographic descriptions of American consumers. VALS breaks consumers into eight different groups based on resources and motivation. For example, one of the groups, the "achievers" are described by VALS in the following way:

> Motivated by the desire for achievement, Achievers have goal-oriented lifestyles and a deep commitment to career and family. Their social lives reflect this focus and are structured around family, their place of worship, and work. Achievers live conventional

lives, are politically conservative, and respect authority and the status quo. They value consensus, predictability, and stability over risk, intimacy, and self-discovery. With many wants and needs, Achievers are active in the consumer marketplace. Image is important to Achievers; they favor established, prestige products and services that demonstrate success to their peers. Because of their busy lives, they are often interested in a variety of time-saving devices.[1]

Technographics is another way to divide and classify target markets. While this categorization is not widely used yet, we imagine with the advent of new technologies, that more will be said about how people use and interact with these items. Technographics categorizes people according to how they use newer technologies such as the Internet, video, cell phones, and so on. The concept was first introduced by Dr. Edward Forrest in a study on VCR usage but has now been adopted by Forrester Research, an IT research and consulting company dealing in marketing and business strategies.

Forrester Research uses what they call the "social technographics ladder" to categorize online consumers into groups based on their usage. The ladder is made up of the following rungs:

- *Creators.* People who make social content (write blogs, upload music and pictures, etc.).
- *Critics.* People who respond to content that others post (write reviews, make comments, or participate in forum discussions).
- *Collectors.* People who organize content for themselves or others by using RSS feeds, tagging, and voting on sites.
- *Joiners.* People who connect in social networks.
- *Spectators.* People who read, watch, and listen but rarely participate in blogs, podcasts, and so on.
- *Inactives.* People who neither create nor consume social content of any kind.[2]

---

1. SRI Consulting Business Intelligence. (2009). *VALS*. Retrieved from http://www.sric-bi .com/VALS

2. Forrest, E. (1988, April/May). Segmenting VCR owners. *Journal of Advertising Research*, *28*(2), 29–39.

Six Basic Questions to Ask About Your Market

I keep six honest serving-men
(They taught me all I knew);
Their names are What and Why and When
And How and Where and Who.

—Rudyard Kipling, "The Elephant's Child," *Just So Stories*, 1902

There is nothing new under the sun, or so it seems. Kipling's century-old poem outlines the six basic questions that everyone from reporters to police detectives still ask when trying to get to the bottom of things. Marketers looking to communicate with potential customers are wise to ask these questions themselves. The answers should be the basis of an efficient, cost-effective, and relevant communications plan.

Those six one-word questions, if addressed critically from a marketing point of view, should open your eyes to more effective communication. You'll find that concentrating on these consumer questions and debating them with your coworkers will be both fun and enlightening.

Who

Who is your market? If you are a retailer, your store may attract specific types of buyers. How well can you define who those buyers are? If you are a manufacturer who markets through distribution channels, there may be more than one "who." There may be the wholesaler, the retailer and the end-consumer, with maybe a key influential thrown in for good measure. In that case, each of them may be different in age, income, gender, needs, and motivation. The more carefully you define your target consumers, the more you can adjust your merchandise to their needs and reach them efficiently. The key is to really study *who* they are because the more you know the better you'll do.

What

What do your customers need to know about you, your products, your hours, and so on? Think about the questions you have been asked over the years by customers and vendors, then see how you answer them. Do your employees know the answers? Look at your communications materials—your Web site, labels, brochures, and so on—and make sure that they address commonly asked questions.

Where

Fish where the fish are biting. Knowing where your buyers are concentrated is the key to cost-effective communications. Focused communications save you money. It makes sense that if your buyers read a certain magazine, that's the place to advertise. It also makes sense to attend consumer shows or industry tradeshows if customers are going to congregate there. The same holds true at the point of sale. Label your products well because potential customers are found at the retail locations that specialize in what you are selling.

When

The question of when to talk to your customers is, like the "where" question, critical to cost-effective communication. You don't want to be communicating when your potential customers are not listening. Ask yourself when does it make the most sense to be communicating with the market. Often, it's a concentrated time period, so you need to be prepared ahead of time for the rush of information you'll be distributing. You might not get a chance to communicate at the prime time for another year, so make the window effective. If you advertise in trade journals or consumer publications, be sure you run several ads in a row building up to your prime season. Research shows that multiple insertions build up readership and awareness.

How

Sometimes the communication vehicle is a well-designed label. Sometimes it's an e-mail message, a social media ad, a Web site, direct mailer, magazine, or newspaper ad. Knowing the answers to the previous four questions makes it easier to answer how to reach your customers most efficiently.

Why

Why should your customers buy from you rather than one of your competitors? What is the benefit to them of buying your products? Ask these questions and if you can't come up with some good answers, you have some brand-building work to do. If you can answer them, you need to make sure that you are communicating why your merchandise is the right product for them. Remember, one of the keys to successful communication in this overcommunicated world of ours is to sell benefits, not features.

Keep these six questions in mind whenever you create any piece of communication aimed at your customers and vendors. They'll help you stay focused and on track and in the long run they'll make and save you money.

Conclusion

We started this book with three objectives. We wanted to introduce the brand map, create an easy-to-read book, and simplify a subject that has had millions of words written about it. Hopefully, you found the brand map to be an easy visual guide to the complex world of branding. This book was intended as a quick read. The tips and methods included were to give you an idea about how you can apply the principles outlined in the book. We don't for a minute doubt that there are other ways to accomplish the same ends. In fact, we know there are. So don't be afraid to use your own method or modify ours if it takes you where you want to go.

We've talked about relationships as being the key to successful branding. Another word we use is connection. Those two words can be considered one and the same because a relationship is a connection between one person and another person, a person an animal, or even a thing. Connections are a two-way street. You give and you get. If you, as a business-person, are giving something to people, be that great service, friendliness, personal attention, or whatever, you'll get loyalty in return.

The key to building successful relationships isn't having a sparkling personality or being smarter than the next person. The key is consistency. Don't let your customers or clients down. Treat them well, consistently. They'll keep coming back and they'll tell their friends about you.

We invite you to contact us to share your thoughts and insights. Over our combined 50-plus years in marketing, we have read scores of books, numerous articles, and discussed brands with hundreds of people. We feel as if we have heard it all and have encountered many examples that we were unable to include in this edition of the book. We're happy to share our experiences with you.

Thank you for reading the book. We sincerely hope that it more than pays for itself. Now, go out and build yourself a brand.

Appendix

Below is a sample **brand manual** table of contents. You can extend each of these items out as far as necessary, add items relevant to your operation, or delete irrelevant items.

1. Introduction (This should be from the CEO explaining how important branding is and how each employee is expected to partake in it.)
2. Brand terminology
3. The brand map
4. Our brand foundation
5. Marketing communications

   - Brand identity standards

     - Naming
     - Logo
     - Color
     - Grid
     - Typography
     - Tagline
     - Iconography

   - Advertising

     - Radio
     - Television
     - Print
     - Social media
     - Out-of-home
     - Web site

   - Collateral

     - Public relations
     - Public speaking
     - Media relations

- Events
- Key messages
- Community involvement
  - Charitable giving
  - Sponsorships
  - Industry associations
  - Community boards
- Style
  - Tone
  - Keywords
  - Voice
- Trade shows

6. Our products and services (For each product and service, address the seven points below.)

- Product design
- Service design
- Packaging
- Retail products
- Sensory equity
- Comarketing
- Brand extensions

7. Operations

- Facilities
- Hiring
- Training
- Procedures
  - Complaint handling
  - New client intake
  - Sales process
  - Customer experience

Index

Note: The italicized *f* following page numbers refers to figures.

CPSIA information can be obtained at www.ICGtesting.com
Printed in the USA
243187LV00005B/70/P